AMMONITES

AND

THE OTHER CEPHALOPODS

OF THE PIERRE SEAWAY

Author Neal Larson collecting a nicely sized *Placenticeras meeki*. Neal operates Black Hills Institute of Geological Research, Inc., in Hill City, South Dakota, along with partners Pete Larson and Robert Farrar.

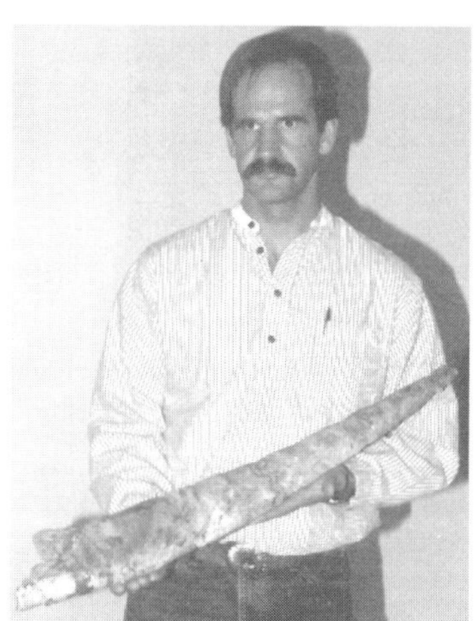

Author Steve Jorgensen with a very complete *Baculites grandis*. Steve is a Consulting Geologist based in Sioux Falls, South Dakota.

Cover: *Placenticeras meeki* from the *Baculites compressus* Range Zone, Pierre Shale, Meade County, South Dakota. Photograph by Ed Gerken.

AMMONITES

AND
THE OTHER CEPHALOPODS
OF THE PIERRE SEAWAY
IDENTIFICATION GUIDE

Authors

Neal L. Larson
Steven D. Jorgensen
Robert A. Farrar
Peter L. Larson

Photographer

Ed Gerken

Exiteloceras

Oxybeloceras

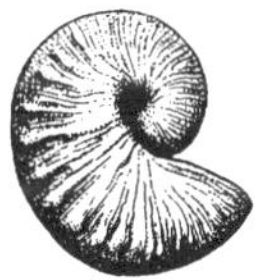

Hoploscaphites

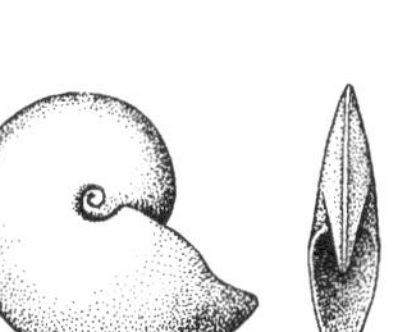

Placenticeras

Baculites

Didymoceras

Pierre Seaway Ammonite Forms

GEOSCIENCE PRESS, Inc.
Tucson, Arizona

Published by Geoscience Press, Inc.
P.O. Box 42948
Tucson, Arizona 85733-2948

Printed in the United States of America
10 9 8 7 6 5 4 3 2 1

Publisher's Cataloging in Publication
   (Prepared by Quality Books, Inc.)

Ammonites and the other cephalopods of the Pierre Seaway : identification
   guide / Neal L. Larson, [et al.].
      p. cm.
      Includes bibliographical references and index.
      ISBN 0-945005-25-3

   1. Cephalopoda, Fossil. 2. Paleontology--Cretaceous. 3.
Paleontology--North America.   I. Larson, Neal L.

   QE806.C47 1997   564'.5'097
                                     QBI96-40453

Distributed to the trade by Mountain Press Publishing Company
P.O. Box 2399, Missoula, Montana 59806
1-800-234-5308

DEDICATION

The authors dedicate this book to Dr. W. A. (Bill) Cobban, researcher, geologist, and scientist with the U.S. Geological Survey in Denver, Colorado, since 1948. Bill's interest in ammonites began as a high school student collecting near Great Falls and Shelby, Montana, and continues to the present day. Many well-deserved honors have been bestowed on Bill over the years. They include: the Meritorious Service Award (1974) and the Distinguished Service Award (1986), from the U.S. Department of the Interior; election as Fellow of the American Association for the Advancement of Science (1982); the Distinguished Geologist Pioneer Award (1985) from the Rocky Mountain Section of the Society of Economic Paleontologists and Mineralogists; the Paleontological Medal from the Paleontological Society (1985); and the Raymond C. Moore Paleontology Medal awarded by the Society of Economic Paleontologists and Mineralogists (1990).

Bill's approach of stratigraphy through paleontology led to his organizing more than 70 different Cretaceous ammonite horizons and his describing and naming over 100 new ammonite species from the Western Interior. Dr. William Cobban probably knows more about the Western Interior Cretaceous rocks and fossils, especially the Pierre Shale, than any other individual in the world.

Yet it is not only ammonites from the Western Interior that Bill has studied and described. Ammonites from the Atlantic, Gulf, and Pacific regions as well as ammonites from Africa, South America, Europe, and Asia have been objects of his research and subjects of his papers. Due to this wide-ranging research and the excellence of his work, 12 species and several genera of Jurassic and Cretaceous invertebrate fossils have been named for Bill by his colleagues in the United States, France, Japan, and Russia.

Bill has consistently and unselfishly shared his knowledge with all interested collectors and researchers. In the process, he has taught skills and shared information that was not available at any university anywhere in the world. It is due, primarily, to Bill's encouragement and assistance that this manuscript was attempted and finally completed. Therefore, on behalf of all the amateur, professional, and academic geologists and paleontologists who have been enriched by Bill's efforts, I say thank you. Bill, this book is for you!

Dr. William A. (Bill) Cobban

Neal L. Larson, Vice President
Black Hills Institute of Geological Research, Inc.

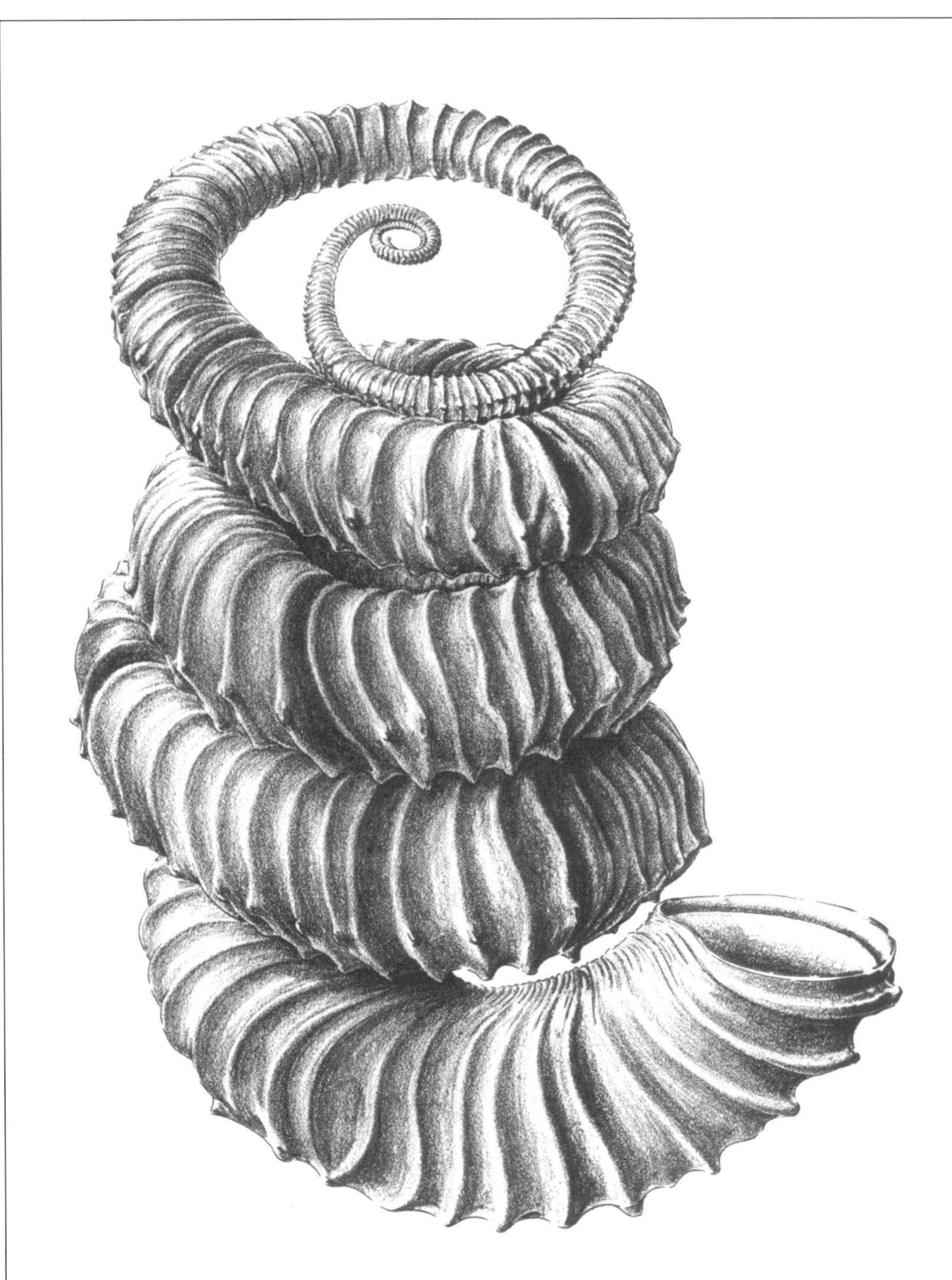

Didymoceras binodosum
Drawing by John R. Stacy

CONTENTS

FOREWORD

The identification of ammonites and other cephalopod fossils has been a tremendous, almost overwhelming challenge for several reasons. First, there are numerous species of cephalopods that appear to be very similar to each other. Many species have not yet been described, named, or properly placed among related forms. Second, the existing descriptions are scattered through a vast array of literature, often in old, out-of-print publications, government agency or museum special publications, or otherwise not generally available. Also, there are relatively few experts to whom one may turn for help. Fossil cephalopods have attracted human interest for centuries and remain among the most popular fossils today, yet there are few people who really know very much about them.

This publication makes a quantum leap forward in solving some of these problems. It brings together, for the first time, the descriptions and references for all of the known cephalopod fossils representing a major geographical area and time span in North America, the Pierre Seaway that divided the continent during the Cretaceous period. This work immediately solves the problem of the scattered literature by bringing these references and descriptions together under one cover. However, it also contributes to solutions of the other two problems. Having all recognized species in one reference will greatly facilitate the naming and description of yet unknown forms. It will also make it possible for many people to become familiar with and knowledgeable about this fascinating group of organisms. Perhaps persons who formerly only admired these fossils for their beauty will now be able to understand more about them and, hence, appreciate them on an entirely new level.

The authors of this publication are among the few experts in the subject and are eminently qualified to write this book. They have worked closely with many if not most of the other authorities in the field; they have assembled their own library of all of the pertinent literature; and, through their own extensive collections of fossil cephalopods and collaboration with other collectors and collections, they have direct access to a major portion of the best fossil cephalopod specimens available anywhere.

Furthermore, these authors know how to help other people. They have not only spent innumerable hours trying to understand ammonites and other fossils to increase their own knowledge, but they also invest an inordinate amount of time and effort in helping others learn, ranging from elementary school children and casually interested collectors to university students, serious amateurs, and fellow professionals. From the first day we met them, my wife and I were graciously offered their assistance in our understanding of ammonites. As a result, through the ensuing years, all the authors became close personal friends and are among our favorite colleagues. I and others from our university have taken numerous groups of students to the Black Hills Institute of Geological Research for tours and education. While visiting the Black Hills Institute, I have been frequently impressed at the numbers of school children, people from the general public, upper-level students, and professional paleontologists who come through their facility. Institute staff willingly, even eagerly, spend many hours explaining their favorite subject—fossils. This publication is an extension, in print, of that outreach.

James W. Grier, Professor of Zoology
North Dakota State University
Fargo, North Dakota

PREFACE

This book began as what appeared to be a fairly simple project in July of 1993, consolidating into one reference the descriptive work on ammonites of the Pierre Shale. Yet even simple concepts can become incredibly complex as you begin to explore them, and thus complicate your work and your life. For example, I wanted to use everyday language to describe the fossils in this book. It quickly became evident that using everyday language would often entail using 20 or more words instead of the one definitive scientific term. Consequently, when expedient, I chose to use the scientific terms. That necessitated compiling a glossary, containing definitions of the common morphological and geological terms used in the scientific description of cephalopods, for inclusion in the book. In the beginning, I planned to define the 50 or so described ammonite species and assumed the writing would take one to three months. My research, however, revealed nearly 100 described ammonite species. Who knew?

This project took hundreds of office hours, consuming a majority of Ed Gerken's, Bob Farrar's, Marion Zenker's, and my scheduled work hours. There were also many evenings and weekend hours invested that kept me from my children and my wife, Brenda. And, as the book grew, so did my family. In October of 1994, our daughter, Elisha, joined her four brothers in our family.

Though that was not its purpose, my immersion in this project has, at times, allowed my mind to momentarily escape the less pleasant ordeals, related to a dinosaur named "Sue," that have followed us over the past few years.

After several years of research and writing, with Bill Cobban's invaluable help, we had nearly completed the text when Robert (Bob) Farrar suggested that we include the remaining six or so cephalopods from the Pierre Seaway. I consented, assuming it would be a three- or four-day project. Many months later we were still discovering and gathering new information about these few species. One consequence of this addition was an opportunity to visit with Dr. Curt Teichert (deceased), who completed *Part K, Treatise on the Mollusca 3* for the Geological Society of America in 1964. Although Dr. Teichert was retired, in very poor health, and confined to a wheelchair, he managed to visit his old office at Harvard University and rummage through his 30-year-old files to locate the Agassiz 1847 reference for the Family Nautilidae for us. To our knowledge, this is the only publication about *nautilus* that contains this reference in its bibliography. As we go to press, these remaining cephalopod descriptions and discoveries have become a major contribution to this work.

During the compiling of the information for this book, we were fortunate to have many cephalopod and mollusc collectors and researchers from throughout North America visit Black Hills Institute. Each person contributed important information, ideas, and data to this manuscript. Now, three years later, the end of this project finally approaches. Even as we ready the book for publication, it is clear that within just a few years, it will need revision and updating.

For me, the most interesting aspect of this project has been the research. It was when we began to match the ammonite descriptions with the fossil ammonites, though, that the true purpose of this work became really clear. We learned quickly not to accept all scientific literature as fact, because discrepancies were found in many scientific ammonite descriptions. Other consequences of our research have been the discovery of more than a dozen new cephalopod species, the elimination of several existing species names, and the serious questioning of at least one currently accepted ammonite genus.

The purpose of this work was to assemble descriptions of the more common ammonites into one consolidated reference, as well as to combine scattered source material, and create a comprehensive, bibliographic and faunal listing for the Western Interior Pierre Seaway. Not all of the ammonites are herein described, because after nearly 150 years of collecting and research on the cephalopods from the Pierre Seaway, there are still many yet to be defined.

The number of cephalopod specimens that have been collected from the Pierre Shale and dispersed throughout the world must be staggering. Most popular localities for collecting invertebrates within the Pierre Shale produce thousands of ammonite specimens every year. With the number of collectors who visit these localities annually and the fact that many of the sites have been collected continuously for more than 100 years, you can begin to understand why these ammonites are not rare, even though good ammonites might be uncommon.

Nearly every rock and mineral collection in this country, not to mention the amateur and museum fossil collections, have at least one ammonite from the Pierre Shale among their specimens. Most of these specimens are misidentified or remain unlabeled. Whether the ammonite is a scaphite, baculite, or placenticeras, most people have had no available source to consult for proper identification.

Although there are hundreds of scientific books and papers written on the subject, very few are written for the amateur, or even the museum curator, that can help them make accurate and scientific identification of their collections.

This book was written with the help and advice of the top ammonite experts in this country. They read, reviewed, and commented on this manuscript throughout its many stages, from its creation to what you now hold. We sincerely hope that this book will serve as an indispensable identification guide for you.

The specimens depicted in this book were primarily from two sources: The U.S. Geological Survey (USGS), Denver, Colorado, and the Black Hills Institute of Geological Research, Inc. (BHI), Hill City, South Dakota.

The USGS specimens are designated by USNM numbers and are part of the U.S. National Museum's permanent collection. They were photographed with the permission and under the supervision of Dr. W. A. Cobban of the USGS. The Black Hills Institute of Geological Research specimens are designated by BHI numbers and are part of that permanent collection. Some of these specimens may be seen on exhibit at the Black Hills Museum of Natural History in Hill City, South Dakota. The remaining photographs of specimens were used with the permission of the owners and are acknowledged with the photos.

This project was fun and very educational for the authors. Our hope is that collectors and researchers will also find it interesting and helpful for many years to come. The authors welcome any suggestions, thoughts, or criticisms of this book that may help make it more enjoyable, useful, or readable.

Neal L. Larson

Anaklinoceras n. sp.
Drawing by John R. Stacy

ACKNOWLEDGMENTS

We thank Dr. W. A. (Bill) Cobban, Dr. Neil Landman, and Dr. Karl Waage for reviewing this manuscript and adding their valuable suggestions and comments to the text. Thanks also to Dorothy Sigler Norton for her incredible artwork and Ed Gerken for his wonderful photography and his many hours on the computer formatting, typing, and inserting the images .

In addition, the authors thank the U.S. Geological Survey for its long-term commitment to the study of the North American Western Interior Pierre Seaway during the last 140-plus years. Their findings have been made public in papers, bulletins, monographs, and maps. Of this esteemed group of geologists, none has been more helpful than Dr. William (Bill) Cobban of the U.S. Geological Survey, who for over 20 years has shared his knowledge, wisdom, and friendship with us. Other scientists who unselfishly shared their time, advice, and wisdom are: Dr. Karl Waage of the Yale Peabody Museum, New Haven, Connecticut; Dr. Neil Landman of the American Museum of Natural History, New York; Dr. Karl Hirsch (deceased) of the University of Colorado at Boulder; Dr. William James (Jim) Kennedy of University of Oxford Museum, England; and Dr. Jim and Joyce Grier from North Dakota State University in Fargo.

In addition, we thank all the ranchers who opened their lands to us and others for collecting, as well as the numerous collectors who have shared their finds with us.

Special thanks go to David Anderson, George Burg, Dr. Ray A., Dorothy, and Japheth Boyce, Earl Brockelsby (deceased), Vern Brooks, Barry Brown, Ed and Ava Cole, Howard (Bud) Ehrle, Steve Haire, Rev. Harry Heidt, Lonnie Holsworth (deceased), James A. Honert, John Larson, Neal C. and Gert Larson, Jim Michaud, Dorothy and Richard Norton, Wayne Olson (deceased), Don Parsons, Paul (deceased) and Winifred Reutter, Bill and Jean Roberts (both deceased), Jordan Sawdo, Jim Schoon, Dean and Donna Talty, Tom Trask, Leon Theisen, Tom Wooden, and Al (deceased) and June Zeitner. These are the people who opened the doors, showed us the way, donated specimens for our research, supported our interest, and even our obsession with these wonderful creatures.

Thanks also to Joe Small, editor of the *Fossil News*, a monthly journal of amateur paleontology, for all of his help on the layout and suggestions to improve readability.

A very special thank you to Brenda Larson, who put up with Neal through all of the hours, weeks, and years that went into this book, when he could often think of nothing else.

And finally, particular thanks is given to Marion Zenker, who typed this manuscript many, many times, created numerous diagrams and drawings, and also suffered the erratic temperaments of the authors with grace. Thank you!

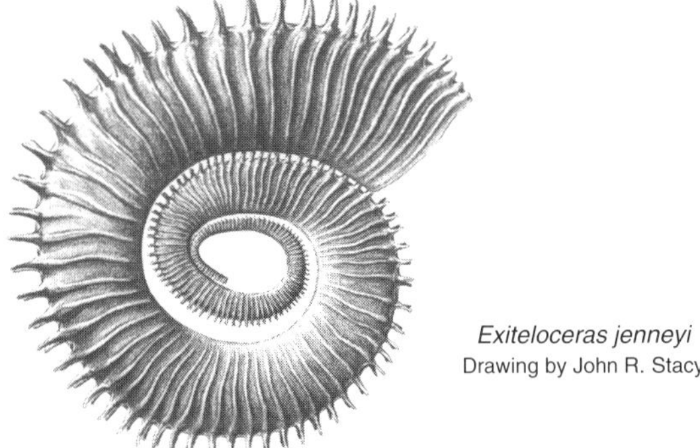

Exiteloceras jenneyi
Drawing by John R. Stacy

Neal L. Larson
Steven D. Jorgensen
Robert A. Farrar
Peter L. Larson

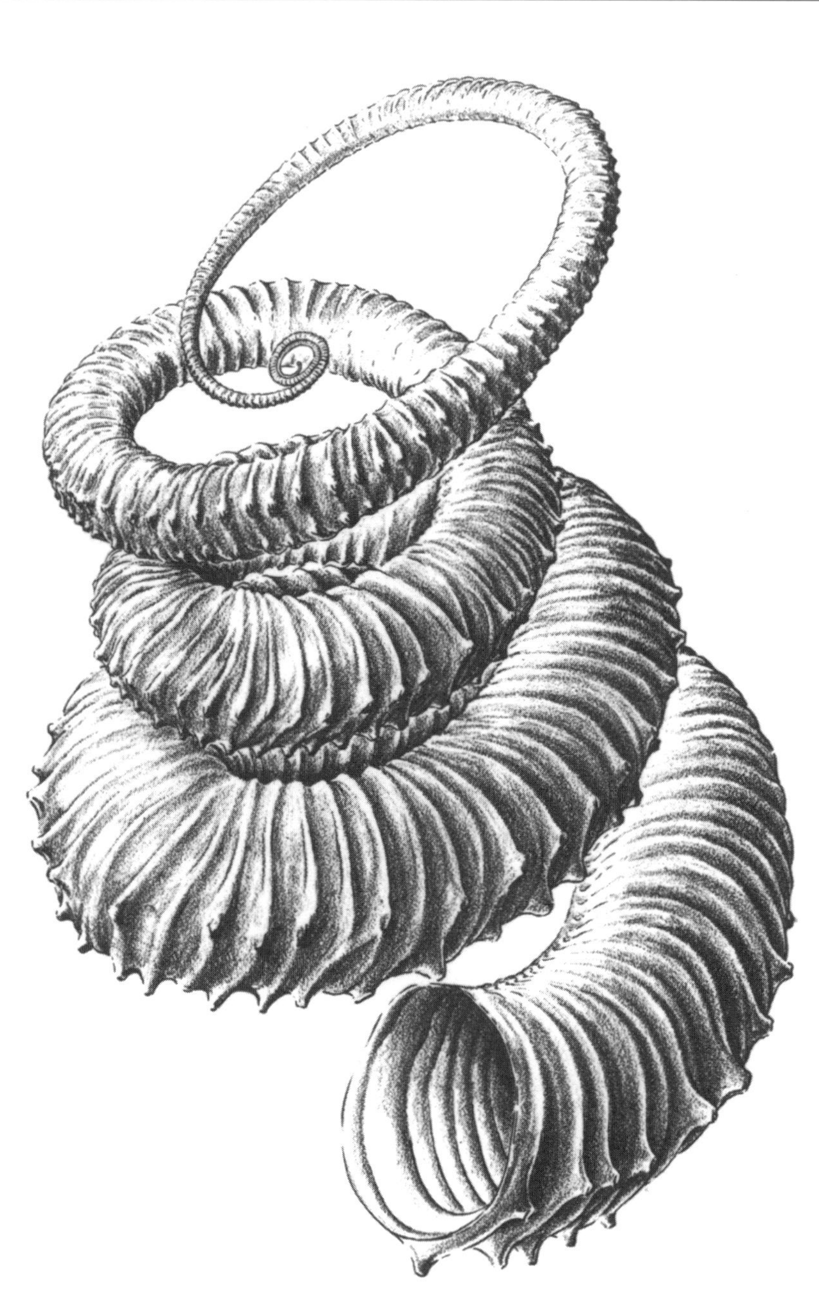

Didymoceras cochleatum
Drawing by John R. Stacy

Introduction

Placenticeras intercalare
(after Meek, 1876)

The Pierre Shale and its macrofauna have been studied for more than 150 years. F. B. Meek collected, figured, and described many of the common invertebrate fossils. His 1876 monograph, *Invertebrate Cretaceous and Tertiary Fossils of the Upper Missouri Country,* is the most comprehensive reference for Pierre Shale invertebrate fossils, to date. Since that time, many others have published works on the cephalopods, gastropods, pelecypods, arthropods, and other invertebrates. A large variety of vertebrates has also been reported. Despite scores of publications on the geology and fauna, many specimens await description. The rapid erosion of the shale exposes large numbers of fossils each year, and the continued collecting of this material will undoubtedly yield new taxa.

Geologic Setting

During the Campanian and Maastrichtian Stages of the Late Cretaceous, the Western Interior epicontinental sea covered an elongated, asymmetrical trough. The western boundary of this Cretaceous basin extended from western Arizona through western Utah, central Idaho, western Montana, and north into the Canadian Arctic. The tectonically active Sévier Orogenic Belt contributed enormous quantities of coarse-grained, clastic material to the rapidly subsiding western portion of the basin.

The eastern edge of the basin can be traced to western Minnesota and Iowa. The fine-grained sediment, low rates of deposition, and thin stratigraphic units indicate a limited sediment source on the eastern side of the basin. The north end of the seaway was open to the boreal regions of northern Canada and Greenland, while the south end of the seaway was connected to the Texas Gulf Coast area. Marine vertebrates and invertebrates migrated into and out of the Western Interior from both of these regions.

Various researchers have investigated the possible depth of the epicontinental seaway, in general, and the Pierre Seaway specifically. Kauffman (1977) states that the shallow-water marine, marginal marine, and coastal-plain coarse-grained clastics, deposited on the western shoreline, were deposited in water less than 50 m deep.

In western Colorado, New Mexico, and parts of eastern Utah, the Mancos and Lewis Shales appear to have been deposited in deep, quiet water (200–300 m), possibly the deepest part of the seaway. The eastern Colorado and western Kansas "hingeline" represent outer shelf water depths of 100–200 m. The eastern platform, in eastern Kansas, Nebraska, and Iowa, probably was rarely covered by more than 100 meters of water at maximum transgression.

Gill and Cobban (1966a) state that at Red Bird, Wyoming, the Gammon Ferruginous and Sharon Springs Members of the Pierre Shale appear to have been deposited in less than 200 m of water, while subsequent members of the Pierre Shale were deposited in more than 200 m of water. Foraminiferal data at Red Bird, Wyoming, indicate a water depth ranging from 120 m to less than 15 m (Mello, 1969).

Gill and Cobban (1973) present a detailed discussion of the depositional environments of the Pierre Shale and its equivalents in the Dakotas, Wyoming, and Montana. Across this vast area, the lithologies range from nonmarine and nearshore arenaceous clastics to fine-grained illite–montmorillonite marine shales. Bentonite beds, of varying thicknesses, can be traced across hundreds, and sometimes thousands, of square miles in this area. These bentonites are valuable correlation markers in the Western Interior because the tiny sanidine and zircon crystals, which occur in the bentonites, have been used to radiometrically date the various members of the Pierre Shale.

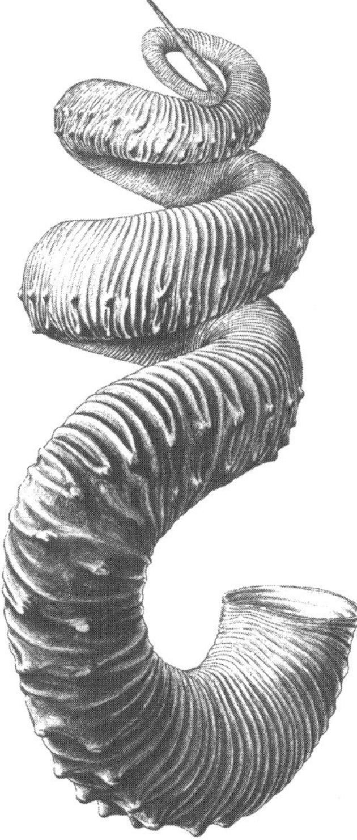

Didymoceras nebrascense
Drawing by John R. Stacy

Position of the Pierre Shale in Geologic Time

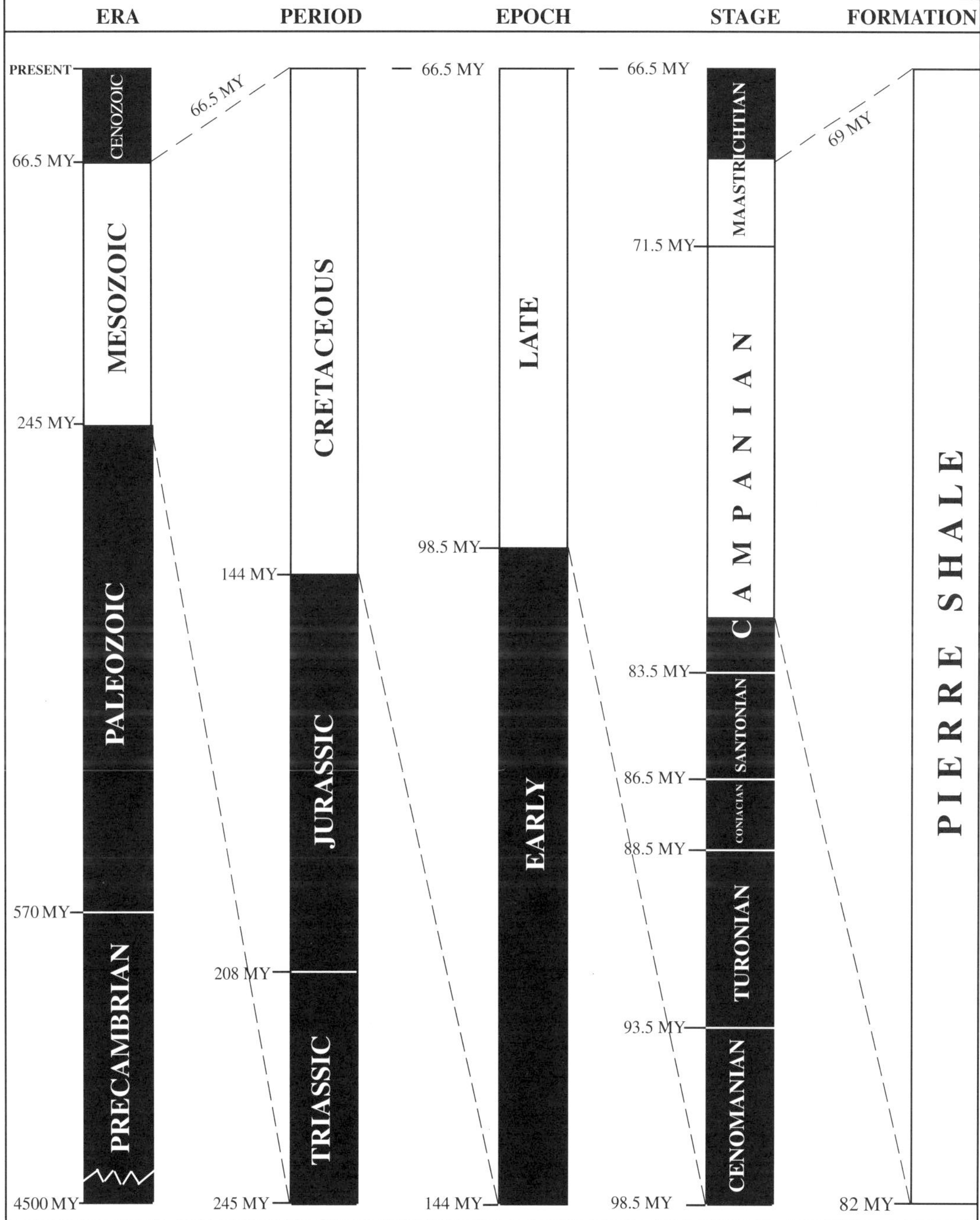

References: Dyman, T. S.; Merewether, E. A.; Molenaar, C. M.; Obradovich, J. D.; Weimer, R. J.; and Bryant, W. A. 1994. Geological Society of America 1983.

Figure: Farrar, N. Larson, 1994
Computer Graphic: M. Zenker

Ever since 1980, when Louis and Walter Alvarez stunned the scientific community with their asteroid impact theory, a search has been underway to locate the crater blasted by the extraterrestrial body hypothesized to have impacted the Earth at the end of the Cretaceous, bringing to an end the reign of the dinosaurs and the ammonites. Two impact features that appear to have been created at or near the end of the Cretaceous include the Chicxulub structure, of the Yucatan Peninsula on the Gulf of Mexico, and the Manson structure in Iowa. The Chicxulub structure is thought to more closely coincide with the Cretaceous–Tertiary (K-T) boundary (64.4 MYA), while the Manson structure is older (73.8 MYA) and occurred during the time of the deposition of the *Didymoceras stevensoni* Range Zone of the Pierre Shale. Izett et al. (1993) reported that the base of the Crow Creek Member of the Pierre Shale in eastern South Dakota contains shocked quartz grains, presumably derived from the Manson impact ejecta, based on radiometric data and proximity.

Possible evidence of the erosive force of a tsunami produced by the Manson impact is the regional, angular unconformity present at the base of the Crow Creek Member of the Pierre Shale. In Lyman County, South Dakota, the unconformity occurs between the Crow Creek Member and the underlying Gregory Member. Farther east, in Yankton County, South Dakota, erosion has completely removed the Gregory Member and the unconformity lies on top of the even older Sharon Springs Member. It appears that the Pierre Shale, below the unconformity at the base of the Crow Creek Member, has been more deeply eroded in southeastern South Dakota (nearer the Manson site) than in more western areas.

Another possible effect of the impact and resulting tsunami is the absence of *Didymoceras stevensoni* and any fossil molluscan or other distinguishable fossils throughout the entire *Didymoceras stevensoni* Range Zone in North Dakota, South Dakota, or Nebraska. As far away as Western South Dakota there is a noticeable absence of fauna between the *Didymoceras nebrascense* Range Zone and the *Extiloceras jenneyi* Range Zone on the northern, eastern, and southern flanks of the Black Hills. The tsunami could have left the seaway uninhabitable or it may have destroyed all evidence of life in these areas. However, the fauna is remarkably intact and abundant on the western and northwestern flanks of the Black Hills, indicating that there must have been some structure that protected or shielded them from the force of the tsunami.

It is geological events such as this that make the investigations of the geology and paleontology of the Pierre Shale interesting, challenging, and exciting. With continued research into the Manson impact structure and associated geology, a more informed and clearer picture will develop that may prove the incredible force of this impact.

Comet Hyakutake
Photo by O. Richard Norton, Science Graphics

THE PIERRE SEAWAY

Throughout the Cretaceous Period, the Western Interior Seaway made a series of transgressions and regressions into the Western Interior Basin of North America. Even though the seaway was open to the Arctic Ocean, there is no evidence in the fossil record to suggest a cold environment. A wide variety of life was supported by three major ecological areas within the seaway—a cool northern boreal zone, a mild central zone, and a warm southern tethian zone.

Ammonite diversity seemed to reach its peak within the temperate zones of the seaway which extended from what is present-day northern New Mexico through southern Canada. Many different types of cephalopods flourished in these waters, but heteromorph ammonites predominated. Ocean warming and cooling plus rapid fluctuations of oxygen content and salinity caused changes in the lithologic deposition and life forms in the seaway. Many extinctions occurred during these episodes, but new forms of ammonites arrived and flourished. The new forms probably came most often from the Gulf Coast, but some probably also arrived via the northern corridor to Greenland or to the Arctic Ocean.

The *Baculites obtusus* Zone, of middle Campanian age, coincides with the maximum expanse of the seaway. Many transgressions and regressions followed the *Baculites obtusus* transgression. The seaway never again covered such a large area, and it finally succumbed to the continental uplift and disappeared near the end of the Maastrichtian.

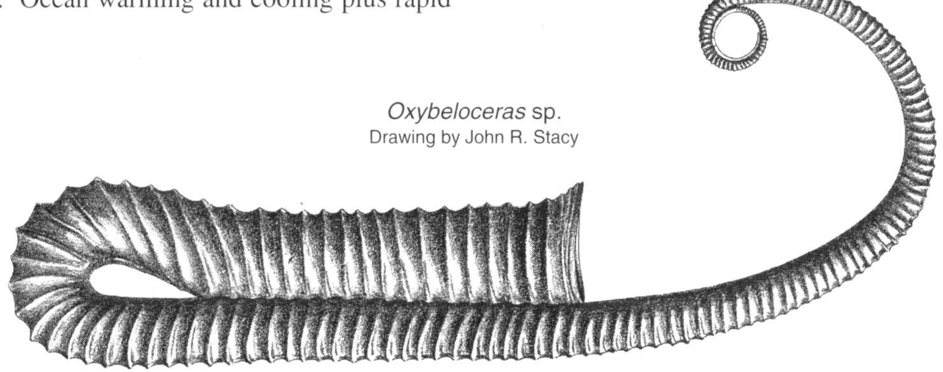

Oxybeloceras sp.
Drawing by John R. Stacy

EXPLANATION OF SUBSEQUENT PIERRE SEAWAY MAPS

Plotting the shorelines of an ancient seaway is difficult and often imprecise work. We are grateful that Dr. William Cobban undertook to plot the west coast of this seaway, using ammonite zones from New Mexico to Canada as his guides. His calculations probably follow quite closely the western shoreline of the Pierre Seaway at its greatest incursion. Glaciation and erosion have nearly obliterated all traces of the seaway's northern and eastern boundaries. A few small outcrops remain, however, suggesting that the eastern edge of the seaway may have extended well into present-day Iowa and Minnesota. Judging from the foraminifera and an ammonite found in glacial till, the northern boundary was probably open to what is now Greenland. Exposures of marine outcrops in the Northwest Territory and the Yukon indicate that at its greatest extent the Pierre Seaway probably opened to the Arctic Ocean to the north.

Pierre Shale Marine Age Outcrops

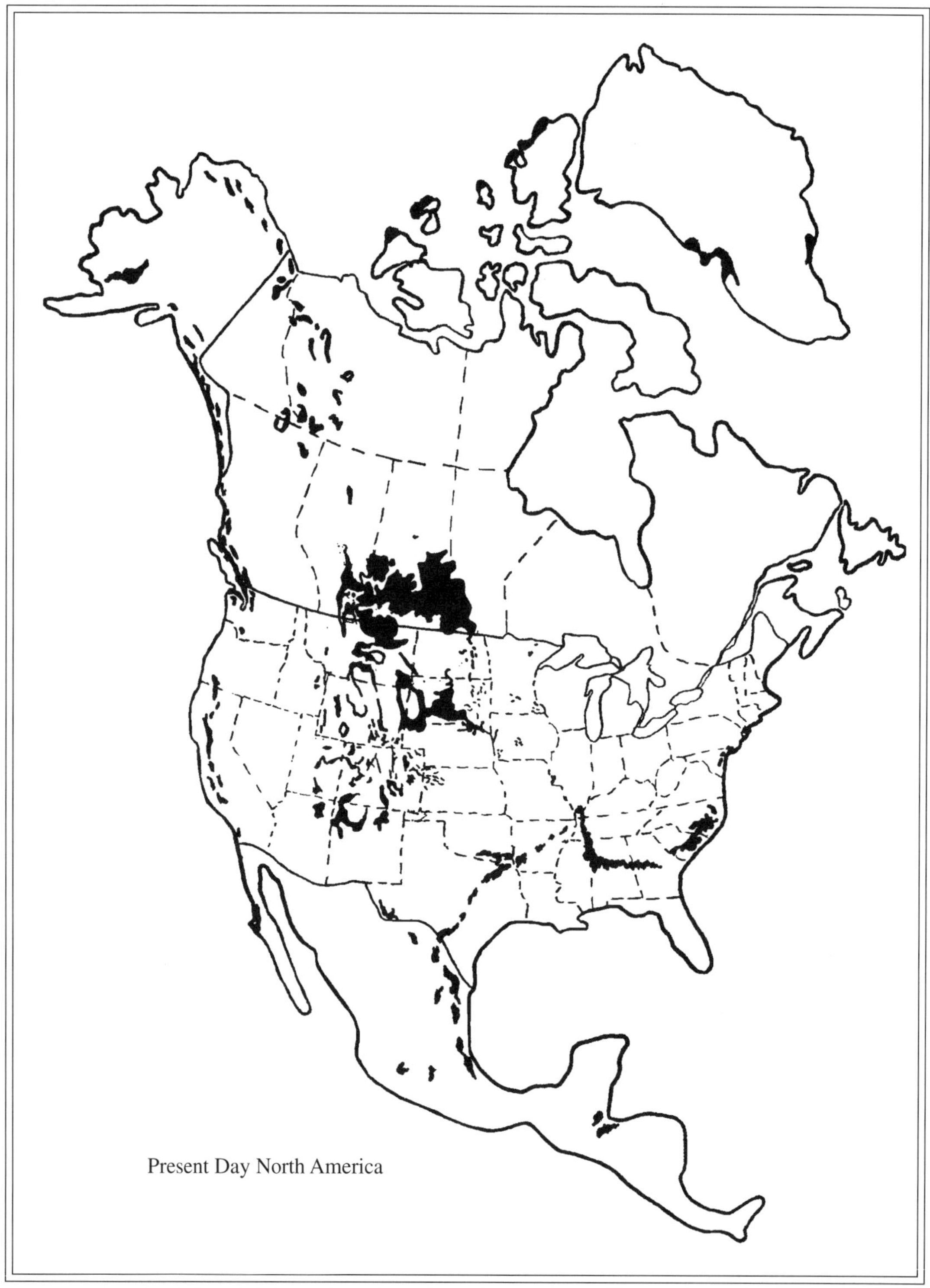

References: U. S. Geological Survey state geological maps; Canadian Geological Survey Provincial geological maps & map of the Arctic; American Association of Petroleum Geologists state geological highway maps.

Figure: N. Larson, 1994
Computer graphics: M. Zenker

Ammonites and the other Cephalopods of the Pierre Seaway

WESTERN INTERIOR PIERRE SEAWAY OF NORTH AMERICA
Regressions and Transgressions during Upper Cretaceous Period

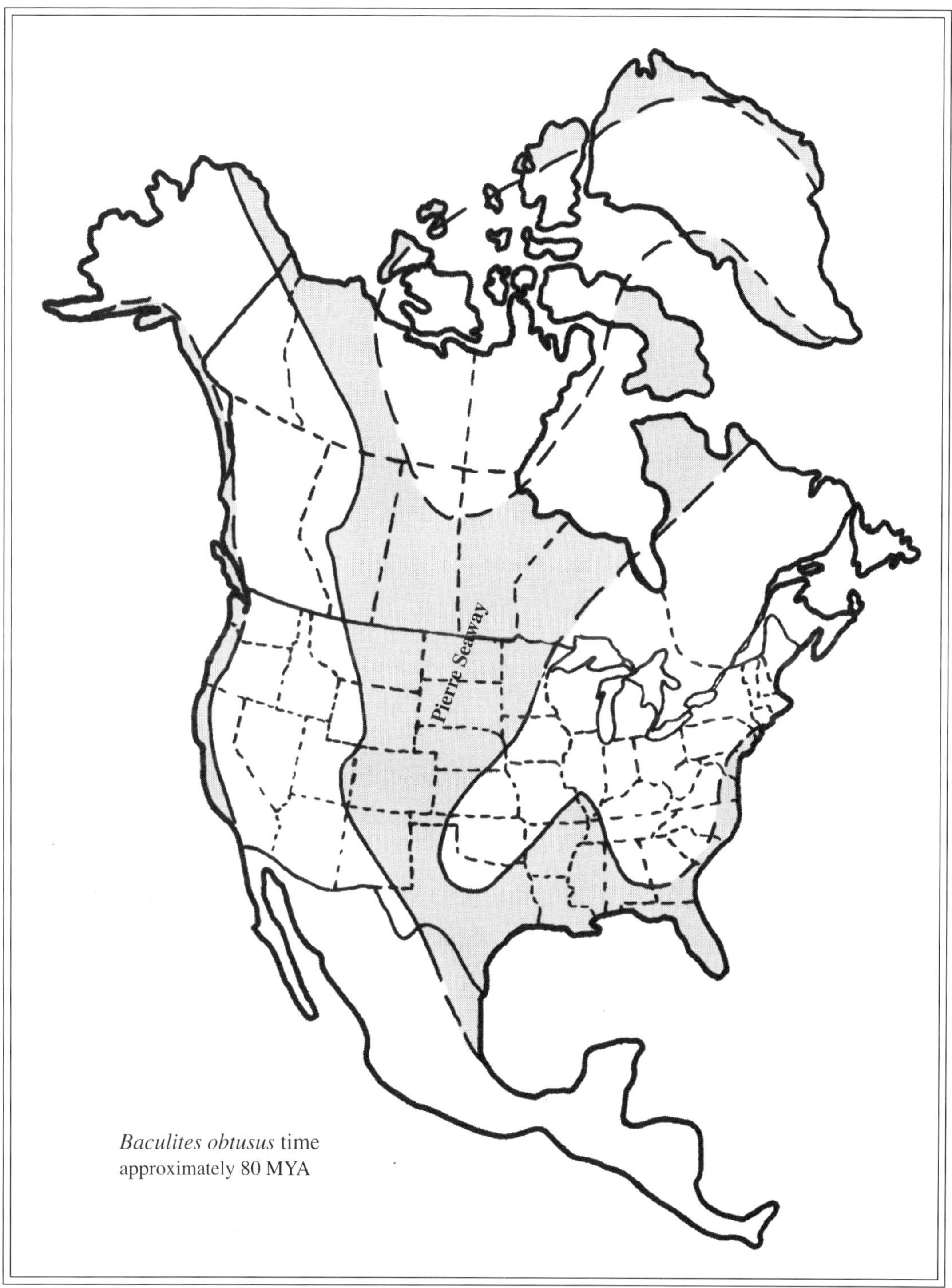

References: Cobban, W.A.; Merewether, E. A.; Fouch, T. D.; and Obradovich, J. D., 1994.

Figure: N. Larson, 1994
Computer graphics: M. Zenker

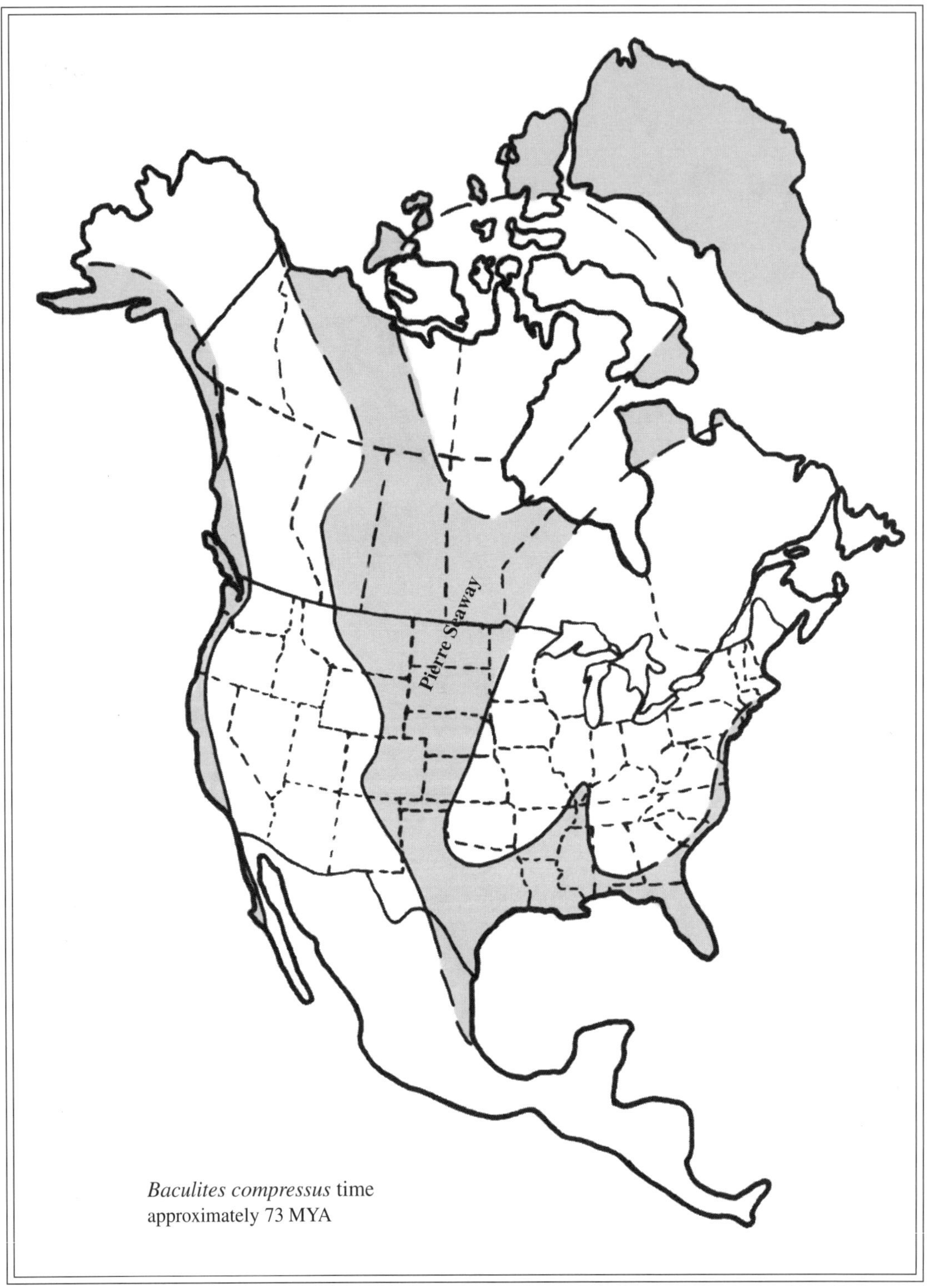

Baculites compressus time
approximately 73 MYA

References: Cobban, W. A.; Merewether, E. A.; Fouch, T. D.; and Obradovich, J. D., 1994.

Figure: N. Larson, 1994
Computer graphics: M. Zenker

WESTERN INTERIOR PIERRE SEAWAY OF NORTH AMERICA

Regressions and Transgressions during Upper Cretaceous Period

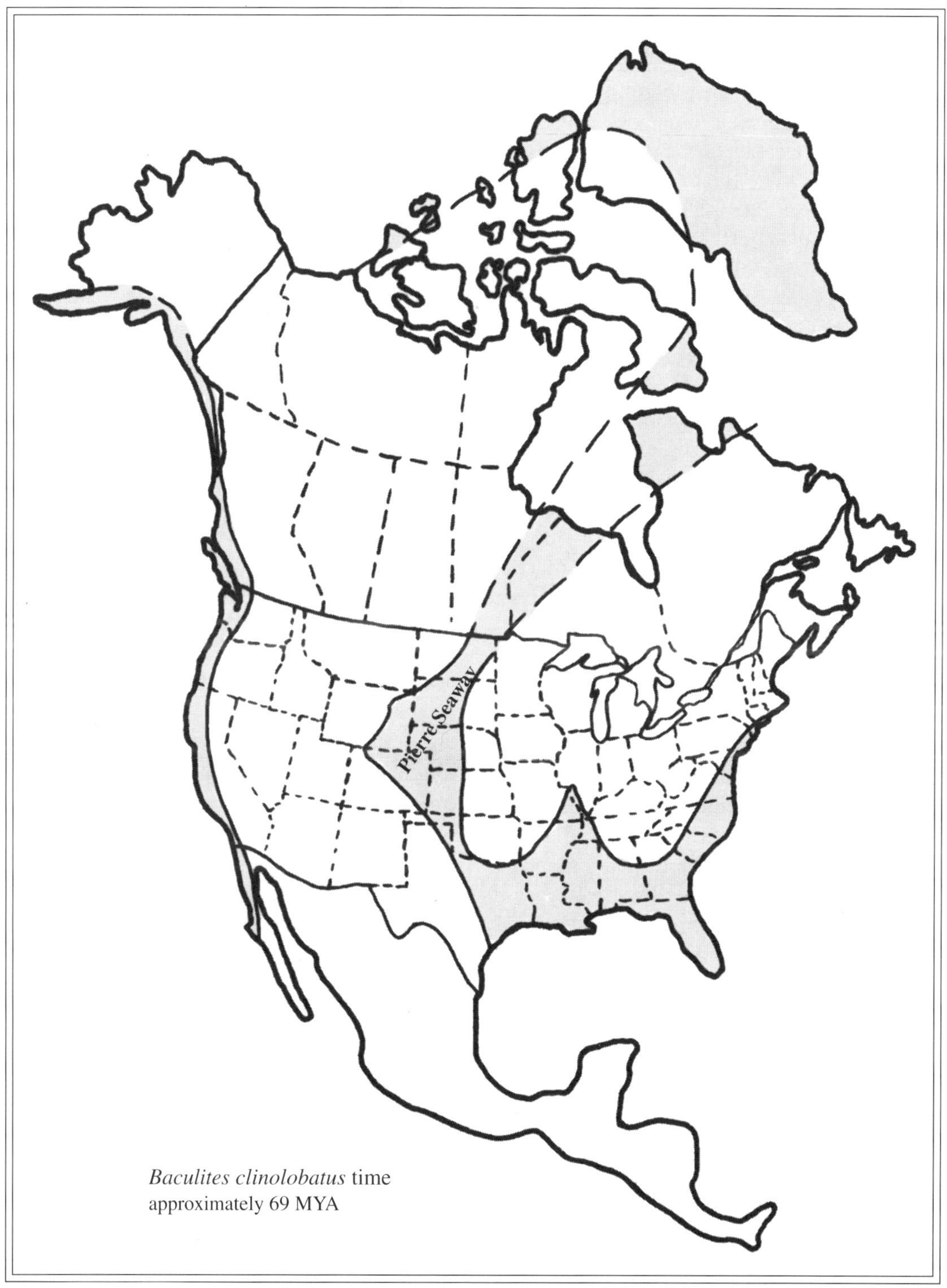

Baculites clinolobatus time
approximately 69 MYA

References: Cobban, W. A.; Merewether, E. A.; Fouch, T. D.; and Obradovich, J. D., 1994.

Figure: N. Larson, 1994
Computer graphics: M. Zenker

The Pierre Shale is largely composed of dark-gray to black marine shale ranging from less than 1,000 to 10,000 ft in thickness. The presence of silty, sandy, and calcareous units within the largely dark-gray Pierre Shale has been the basis of formal member names, such as the Gammon Ferruginous Member (including the Groat Sandstone Bed), Sharon Springs Member, Mitten Black Shale Member, Red Bird Silty Member, and Monument Hill and Kara Bentonitic Members of the Black Hills area; and the Sharon Springs, Gregory, Crow Creek, DeGrey, Verendrye, Virgin Creek, Mobridge, and Elk Butte Members of the Missiouri River valley of South Dakota. Nearshore sandstone units in northern Colorado have given rise to such names as Hygiene, Terry, Larimer, Rocky Ridge, and Richard Sandstone Members of the Pierre Shale. The monotony of the shale units is broken by many horizons of calcareous and ironstone concretions. Many of these concretions contain fossils. Due to the difficulty recognizing structure in the shale, the concretion horizons with distinctive faunal assemblages have been extremely important to geologists working in the Pierre Shale. Throughout the Western Interior, the fossilized remains of ammonites are so numerous and widespread, and their morphological differences so discernible, that by using them, geologists and paleontologists can separate very thick homogeneous rock units into smaller biostratigraphic zones. These limited zones enable geologists to locate themselves at the same point in geologic time across great distances (sometimes more than 1,000 mi apart). Dr. William A. Cobban invested more than 40 years working with the Upper Cretaceous rock units of the Western Interior Seaway and helped define 26 different ammonite range zones in the Pierre Shale and age-equivalent rocks to date. The recognition of these zones has been of major importance in unraveling the history and the stratigraphy of the Pierre Shale. Glenn R. Scott and William A. Cobban mapped these ammonite zones along the Front Range from south central Colorado northward to the Wyoming border.

During Campanian and Maastrichtian times fluctuations in sea level caused changes in the western (and presumably eastern) margins of the seaway. The western margin of the Pierre Shale is often intertongued with terrestrial deposits. One such terrestrial "tongue" is the Campanian Judith River Formation of central Montana and Alberta. During late Campanian time, the seaway again expanded west depositing the Bearpaw Shale of Montana and Alberta. The marine Bearpaw Shale is a time equivalent of the upper part of the Pierre Shale. As the Cretaceous drew to a close, the Pierre Seaway began to recede eastwardly and then southeastwardly. The Pierre Seaway disappeared by middle Maastrichtian time.

The Pierre Shale rests conformably to disconformably upon the Niobrara Formation, a yellow to gray weathering limestone or chalk. Deposited conformably on top of the Pierre Shale is the Fox Hills Formation, which is chiefly sandstone and sandy shale that was deposited at the edge of the seaway as it receded from west to east. The formation is time transgressive so that the top of the Pierre Shale in Wyoming is older than the top of the Pierre Shale in central South Dakota.

MACROFOSSILS

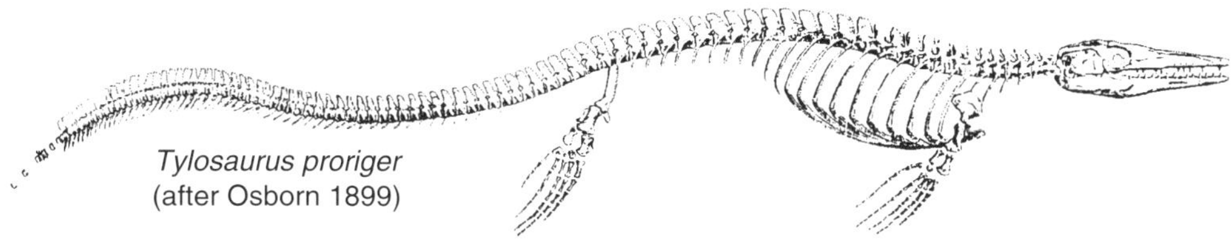

Tylosaurus proriger
(after Osborn 1899)

The Pierre Shale is probably best known for its remarkably well-preserved ammonites. These include the planispiral *Placenticeras* and a host of heteromorph forms such as *Baculites, Hoploscaphites, Jeletzkytes, Oxybeloceras, Solenoceras, Exiteloceras,* and *Didymoceras.* There are also a number of lesser known and several known but undescribed ammonites. In certain horizons, ammonites are found in great numbers and often in an excellent state of preservation with iridescent, nacreous shell material intact. *Eutrephoceras,* probably the ancestor of the present-day chambered nautilus, occurs abundantly in the later portions of the Pierre Shale. Other cephalopods, such as belemnites and squids, are rare. Gastropods, pelecypods, and scaphopods are abundant and are commonly well preserved. Although these mollusks are typically best preserved in calcareous concretions, they also occur loose in the shale.

Echinoderms are rare in the Pierre Shale, although several echinoid genera and a brittle-star have been noted. The echinoids usually occur in concretions but have also been found loose in the shale. One locality in South Dakota has produced complete ophiaroids in compacted shale units.

Arthropods are uncommon in the Pierre Shale; however, limited areas have produced abundant lobsters and crabs. Several localities in South Dakota produce small crabs in concretions (*Dakoticancer*). One site yields crabs and lobsters (*Linuparus*) in concretions and loose in the shale. A concretion horizon in the Bearpaw Shale (equivalent of the upper part of the Pierre Shale) in central Montana produces abundant lobsters of the genera *Hoploparia, Palaeonephrops,* and *Linuparus.* Other invertebrates occasionally recovered include bryozoans, brachiopods, and corals.

Vertebrate fossils are found throughout the Pierre Shale. They are preserved both in concretions and loose in the shale. Vertebrates are common only in the Sharon Springs Member near the base of the formation. Most of the specimens found in this horizon are poorly preserved; consequently, excavation, preparation, and description of this fauna have been limited. Interest in this fauna has increased and extensive collecting projects have begun.

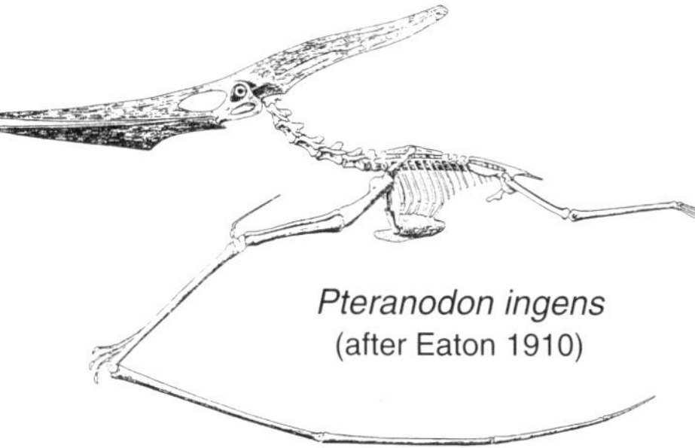

Pteranodon ingens
(after Eaton 1910)

Ken Carpenter of the Denver Museum of Natural History has been particularly active in researching this fauna. Documented vertebrates include mosasaurs, plesiosaurs, pterosaurs, birds, turtles, and fish. Dinosaurs are one of the more unusual faunal elements of the Pierre Shale, probably representing carcasses that were washed out to sea.

AMMONITE ZONES OF THE WESTERN
Pierre Shale and Age

Period	Stage		Ammonite Zones (1)	AGE $^{40}Ar/^{39}Ar$ (2)	South Dakota Wyoming Southern Flank Black Hills (1)	South Dakota Wyoming Montana Northern Flank Black Hills (6)	South Dakota Missouri River (5)	North Dakota Eastern (7)	Nebraska Chadron Arch (5)
UPPER CRETACEOUS	MAASTRICHTIAN	Lower	*Jeletzkytes dorfi*			Hell Creek Formation	Fox Hills Formation		
			Baculites clinolobatus	69.42 ± 0.37	upper unnamed shale		Elk Butte Member	Unnamed Shale Member	
			Baculites grandis			Fox Hills Formation	Mobridge Member		
			Baculites baculus						
			Baculites eliasi		Kara Bentonitic Member		Virgin Creek Member		
				71.3 ± 0.5					
	CAMPANIAN	Upper	*Baculites jenseni*		Lower unnamed shale (part)	unconformity			Upper part
			Baculites reesidei			Unnamed Shale Member	Verendrye Member	Odanah Member	
			Baculites cuneatus		? absent or thin ?				
			Baculites compressus	73.35 ± 0.39	Pierre Shale		?		
			Didymoceras cheyennense			Monument Hill Bentonitic Member	DeGrey Member	?	
			Exiteloceras jenneyi	74.76 ± 0.72	Lower Unnamed Shale Member (part)		Pierre Shale; ?		
			Didymoceras stevensoni			Unnamed Shale Member	Crow Creek Member	DeGrey Member	Pierre Shale
			Didymoceras nebrascense			Pierre Shale	?		
		Middle	*Baculites scotti*					?	Middle part
			Baculites reduncus (3)		Red Bird Silty Member	Red Bird Silty Member	Gregory Member		
			Baculites gregoryensis					Pierre Shale; Gregory Member	
			Baculites perplexus (4)		Mitten Black Shale Member				Mitten Black Shale Member
			Baculites sp. (smooth species)			Mitten Black Shale Member			
			Baculites asperiformis		Sharon Springs Member			Pembina Member	
			Baculites mclearni				Sharon Springs Member		
			Baculites obtusus	80.54 ± 0.55					
		Lower	*Baculites* sp. (weak flank ribs)						Sharon Springs Member
			Baculites sp. (smooth)		Gammon Ferruginous Member	Gammon Ferruginous Member		Niobrara Formation (part)	
			Scaphites hippocrepis III						
			Scaphites hippocrepis II	81.71 ± 0.34			Niobrara Fm. (part)		Niobrara Fm. (part)
			Scaphites hippocrepis I		Niobrara Formation (part)	Niobrara Formation (part)			

1) Gill and Cobban 1966a 2) Obradovich, J. D. 1993

3) The *Baculites reduncus* zone may be the same age as the *Baculites gregoryensis* zone [Cobban written communication 1994]

4) Zone contains an early and late form of *Baculites perplexus* and is separated by *Baculites gilberti*.

5) Gill and Cobban 1972a 6) Cobban and Larson 7) Gill and Cobban 1965

Interior Pierre Seaway
Equivalent Rock Units

Location (reference)	Rock units, top to bottom
Kansas, North Central (8)	Pierre Shale: eroded — ? — Unnamed Member — Salt Creek Member — ? — Lake Creek Member — ? — Weskan Member — Sharon Springs Member; Niobrara Formation (part)
Colorado, Fort Collins (8)	Pierre Shale: Transition Member — sandstone and shale — Richard Ss. Member — Ss. & shale — Larimer Rocky Ridge Ss. Member — sandstone and shale — Terry Ss. Member — shale — Hygiene Sandstone Member — sandstone and shale — Sharon Springs Member — Transition Member; Niobrara Formation (part)
Colorado, Pueblo (8)	Trinidad Sandstone — Unnamed shale; Pierre Shale: Teepee Zone (of Gilbert 1897) — Sego Ss. — Rusty Zone (of Gilbert 1897) — Buck Tongue — Sharon Springs Member — Castlegate Ss. — Apache Creek Ss. Member — Transition Member; Niobrara Formation (part)
Colorado–Utah, Eastern Book Cliffs (12)	Hunter Canyon Formation — ? — Mount Garfield Formation — Sego Ss. — Buck Tongue — Castlegate Ss. — Blackhawk Formation — Mancos Shale
New Mexico–Colorado, Northeastern San Juan Basin (12)	Kirkland Shale — ? — Fruitland Formation — Pictured Cliffs Ss. — Lewis Shale — Mesaverde Formation — Mancos Shale (part)
Wyoming, Central, Salt Creek (1)	Lance Fm. (part) — Fox Hills Fm. — Lewis Shale — Teapot Sandstone Member — unnamed marine shale — Parkman Sandstone — shale — Stray Ss. — shale and bentonite — Sussex Ss. Member — Sandy Shale — Shannon Ss. Member — shale — Fishtooth Sandstone — shale (Mesaverde Formation; Cody Shale, part)
Wyoming, South, Laramie (10)	Lance Formation (part) — Fox Hills Fm. — Lewis Shale — Pine Ridge Sandstone — ? — Rock River Formation — Parkman Sandstone — Steele Shale (part) (Mesaverde Group)
Montana, Central (9)	Hell Creek Formation (part) — Fox Hills Formation — Bearpaw Shale — Judith River Formation — Parkman Sandstone — Claggett Shale — Eagle Sandstone (part) (Montana Group)
Montana, Hardin Area (9)	Fox Hills Formation — Bearpaw Shale — Parkman Sandstone — Claggett Shale — Eagle Sandstone (part) (Montana Group)
Alberta–Saskatchewan, Southern (11)	Eastend Formation — Bearpaw Shale — Oldman Formation — Foremost Formation — Pakowki Shale — Milk River Sandstone
Manitoba, Central (13)	Eastend Formation — unnamed Member — ? — Odanah Member — Millwood Member — Pembina Member — Gammon Ferruginous Member — Niobrara Fm. (part) (Pierre Shale)

8) Gill, Cobban and Schultz 1972a   9) Gill and Cobban 1973   10) Gill, Merewether, and Cobban 1970
11) Cobban 1955   12) Cobban [written communication 1994]   13) McNeil and Caldwell 1981
*Ss is abbreviation for Sandstone

Illustration: W. Cobban and N. Larson.
Computer graphics: M. Zenker, 1996.

CLASS CEPHALOPODA

Cephalopoda are the largest, most intelligent, and most agile representatives of the Phylum Mollusca that have ever lived. Cephalopods (meaning head–foot) have external organs consisting of a distinct head with two large eyes, a beak, a funnel, and a circle of arms with a pair of prehensile tentacles (often with hooks or suckers) around the mouth. They also have a large body cavity that protects the internal organs such as the heart, kidneys, gills, stomach, intestines, reproductive organs, radula, and a mantle cavity. All cephalopods (with the exception of some octopi) have either an external shell where the animal inhabits the last chamber, or an internal shell that is linearly chambered or reduced in size.

Cephalopods have been divided into two groups, Tetrabranchiata (four-gilled), which includes the external shelled nautiloids and ammonoids, and the Dibranchiata (two-gilled), which includes squids, octopi, and the belemnites or cuttlefish. Shelled cephalopods are used by biostratigraphers to separate the thick Paleozoic and Mesozoic rock units into smaller, more concise biostratigraphic zones. Modern cephalopods are most abundant in the shallow (less than 100 m) coastal areas of the world's oceans.

Many cephalopods, as do other predators, have a higher metabolic rate than most other invertebrates. This enables them to be quick, nimble swimmers that can stalk prey and compete with fishes for food. The molluscan foot of the shelled and soft-bodied cephalopods has been modified into a funnel (hyponome) that can be pointed in different directions. When the cephalopod takes water into the mantle cavity, it then expels the water out through the funnel and by jet propulsion moves itself in the opposite direction from the direction in which the funnel points.

Cephalopod brains are so well developed that they are capable of learning and taking care of their young. Many types of cephalopods

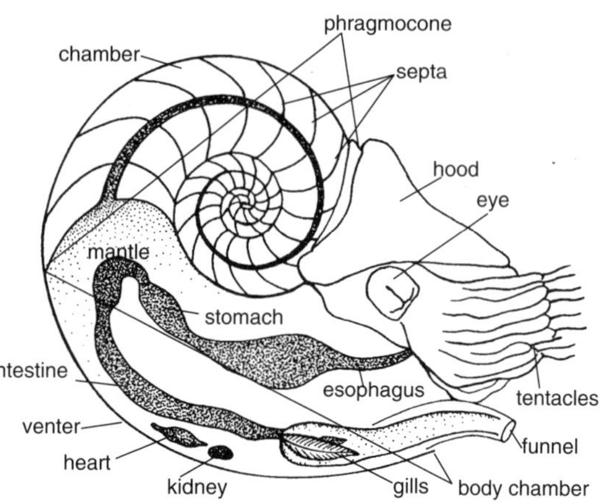

Cross Section of Chambered Nautilus
Illustration by D.S. Norton

display fervent courtship exhibitions. Extinct cephalopods ate a variety of food. The smaller types were primarily plankton eaters, but as with all predatory animals, they were opportunistic feeders, eating whatever they were able to catch. The diet of extant cephalopods consists of plankton, crustaceans, and fish as well as other cephalopods. They, in turn, are fed upon by fish, reptiles, crustaceans, and other cephalopods.

Shelled cephalopods have the ability to achieve neutral buoyancy in water due to their chambered shells which can be filled with gas or cameral fluids to control the depth to which the animal sinks or rises. *Nautilus* (Latin: *nautilos* means sailor), the only surviving, externally shelled cephalopod, even appears in modern fiction. In Jules Verne's 1870 book titled *Twenty Thousand Leagues Under the Sea*, Verne named his submarine, the *Nautilus,* and gave it the characteristics of this remarkable cephalopod.

Shelled cephalopods do not have ink sacs because they have a protective, hard, external, chitinous shell. Lehman (1967) and Wetzel (1969) did describe what they believed to be ink sacs in two different ammonites. Landman and others, however, dispute this (personal communication, 1994). The soft, external-bodied cephalopods typically have ink sacs that enable them to hide from their enemies behind a self-generated ink cloud. The earliest known fossil ink sacs are from the Jurassic. Many modern squid, cuttlefishes, and octopi have the ability to make rapid color changes with the aid of muscle-controlled pigment-bearing cells. This muscle-controlled color changing is unique to cephalopods.

CEPHALOPOD ECOLOGY

Assuming that the main body of the Pierre Seaway ranged in depth from 15 to 200 meters, a vast array of ecological niches existed. Animal life in the sea would have evolved to inhabit the benthic or pelagic environs of the seaway as best suited their particular needs or was easiest for them to exploit. It is possible that the highly buoyant, free-swimming juveniles of all cephalopod genera were dispersed by the surface currents of the seaway while they drifted with the other planktonic organisms living in the sea. This would help explain how species were able to quickly disperse across the seaway. Sub-adults probably migrated to the portion of the water column that they would occupy as adults.

Juvenile cephalopods probably ate other planktonic-sized organisms, both plant and animal, and were likewise consumed by larger creatures. All growth stages of cephalopods

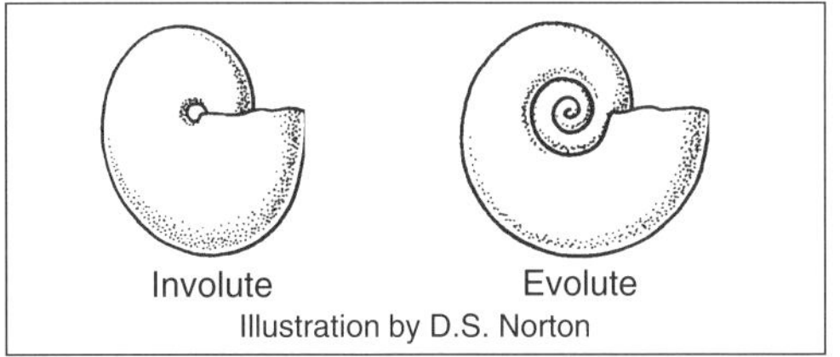

were subject to predation by marine reptiles, birds, fish, crabs, squid, and larger ammonites.

The planispiral genera (e.g., *Menabites*, *Menuites*, *Pachydiscus,* and *Placenticeras*) continued with their involute and evolute, overlapping whorls throughout their lifetimes, but the heteromorphs changed their shapes dramatically.

Scaphites grew involute coils until nearly adult, when the living chamber became retracted from the earlier whorls. *Exiteloceras* was planispiral, but it is still considered a heteromorph because the whorls do not come into contact. *Baculites* began as a coiled ammonitella, but straightened after only about one and one-half coils. All three genera were probably relatively mobile and their neutral buoyancy would have allowed them to actively feed throughout the water column.

Solenoceras began as a typical coiled ammonitella and grew a straight limb that inexplicably turned 180° and grew back in the direction of the ammonitella with the body chamber in contact alongside the phragmocone. The *Oxybeloceras* ammonitella was followed by one and one-half to two open spiral whorls that retracted into a straight to slightly curved limb. Later ontogenetic stages of *Oxybeloceras* look similar to those of *Solenoceras*. Depending on the species, the ammonitella of *Didymoceras* was followed by straight or curved limbs that changed to tight or loose helical coils and ended with a J-shaped or U-shaped body chamber.

Schooling behavior is almost always depicted for *Baculites* as analogous to the modern squid. There is no direct evidence for this behavior in any of these genera. However, the large numbers of individuals of some species would indicate vast schools, not only when spawning, but also when feeding opportunities presented themselves. *Didymoceras, Oxybeloceras,* and *Solenoceras* were not fast swimmers; consequently, they doubtless were benthic feeders, eating plankton, carrion, as well as larger, slow-moving or careless marine animals.

Cretaceous nautiloids probably lived in the same shallow-water depths as the ammonites. Extant nautili are nocturnal and only move up in the water column at night to feed. They prefer to stay at deeper depths during daytime hours, probably to avoid predators. In reef-slope habitats, the nautilus is a bottom-dwelling or benthic scavenger and predator. Modern adult nautili tend to stay in the same area they inhabited as newborns and juveniles; most travel no more than 30 km from their home territory. Long-term habitats appear to be at about 300 m deep or less in the water column, which limits damage to the shell by flooding or implosion.

Squids and octopi are rapid swimmers using their tentacles or arms for catching and holding their prey. They are both benthic and pelagic dwellers that search the reefs and shallow seas for their food. These animals propel themselves by taking water into the mantle cavity and expelling it through an organ called the funnel.

Preserved remains of fossil nautiloids are not nearly as abundant as those of the ammonites. This suggests that they were primarily solitary animals, whereas the ammonites probably "schooled" as evidenced by the abundance of their remains. The population densities of the nautilus were never as large or diverse as the ammonites. This is probably a result of their reproductive success, with a few large eggs laid every year by the female. Due to the fine preservation of cephalopods and other fauna throughout the Pierre Seaway, we have been able to learn a great deal about their habitat.

DIMORPHISM IN CEPHALOPODS

It is important to the paleontologist and biostratigrapher to determine the gender of cephalopods. If male and female differences are not identified, there may be twice as many species described as actually lived because each gender may have dramatically different body size and shell morphology.

The extant *Nautilus* (a living relative of these ancient cephalopods) have been studied for more than 100 years and produce some interesting observations. Collections of specimens of *Nautilus* from New Guinea and the Philippine Islands show that the percentage of males in the mature population ranges from 69% to 92%. This is a surprisingly large percentage of males because it is the female that guarantees the survival of the species. One male can impregnate several females, but the female must wait out the gestation period imposed by nature before producing young.

In mature nautili, the male generally has a slightly larger shell and a heavier visceral mass. Male nautilus shells are larger in diameter and wider than those of the mature female. It is probable that sexual maturity is not reached until maximum growth is achieved. Some giant ammonites, however, surely must have attained sexual maturity long before they reached maximum size.

Another interesting phenomenon of the extant nautilus is that the female lays comparatively few eggs each year. Therefore, even though nautilus has the ability to reproduce for several years, it is probably not capable of explosive reproduction. That may be the primary reason *Eutrephoceras* has never been found in large enough numbers to be designated a Range Zone fossil within the Pierre Seaway.

For more than 100 years, paleontologists assumed that ammonites were sexually dimorphic, but it was not until the 1950s and 1960s that discoveries were made that support this assumption. One of the most exciting finds regarding ammonite sexual dimorphism was the discovery by Lehman in 1966 of preserved egg sacs in the fossilized remains of a Jurassic ammonite macroconch. The macroconch or larger conch of ammonites has since been widely accepted as being the shell of the female. Shell size seems to be the most striking difference between the sexes in ammonites. In the invertebrate world, and in nearly all molusks, females are ordinarily larger than males. There are often strikingly different male and female morphological forms.

In scaphitids, the macroconch (female) has a slight swelling just above the umbilicus in the body chamber. This has been suggested by Landman and others as the possible placement site of a brood chamber. In the Family Scaphitidae, the macroconch body chamber increases gradually in width but rapidly in height.

However, in the microconch, the body chamber grows slowly in both height and width, but never attains the mass of the macroconch. Tubercle placement, flank and venter width, and ribbing are all factors when attempting to establish morphotypes. In the photos of the scaphitids, we show microconchs and macroconchs together so that the reader can see the extreme differences.

The fossil record indicates that the male-to-female ratio in ammonites was probably the opposite of the extant *Nautilus*, thus providing the number of females necessary to make explosive reproduction possible. It is also believed that because of the small ammonitella size in ammonites, they may have laid hundreds or thousands of eggs in a year, which may explain the great numbers of ammonites found preserved in the Western Interior Pierre Seaway. Evidently, water temperature, salinity, depth, and a lack of natural predators made conditions favorable for cephalopods to reproduce rapidly because their fossil record indicates great numbers of these animals inhabited the Pierre Seaway.

If we are to accurately describe these marine animals, it is important to correctly identify the gender of individual specimens. So how does a paleontologist determine the sex of a cephalopod? By studying not only the mature shells but also the phragmocones, and by keeping faunas of specific areas separate for study, we may be able to distinguish sexual dimorphism from species differentiation. It is already known that suture patterns, shell shape, and shell growth are similar in immature forms of a species, regardless of gender. Studying all of these elements in concert will help the scientist assign an accurate gender to a long-dead animal and ensure a more accurate depiction of the species.

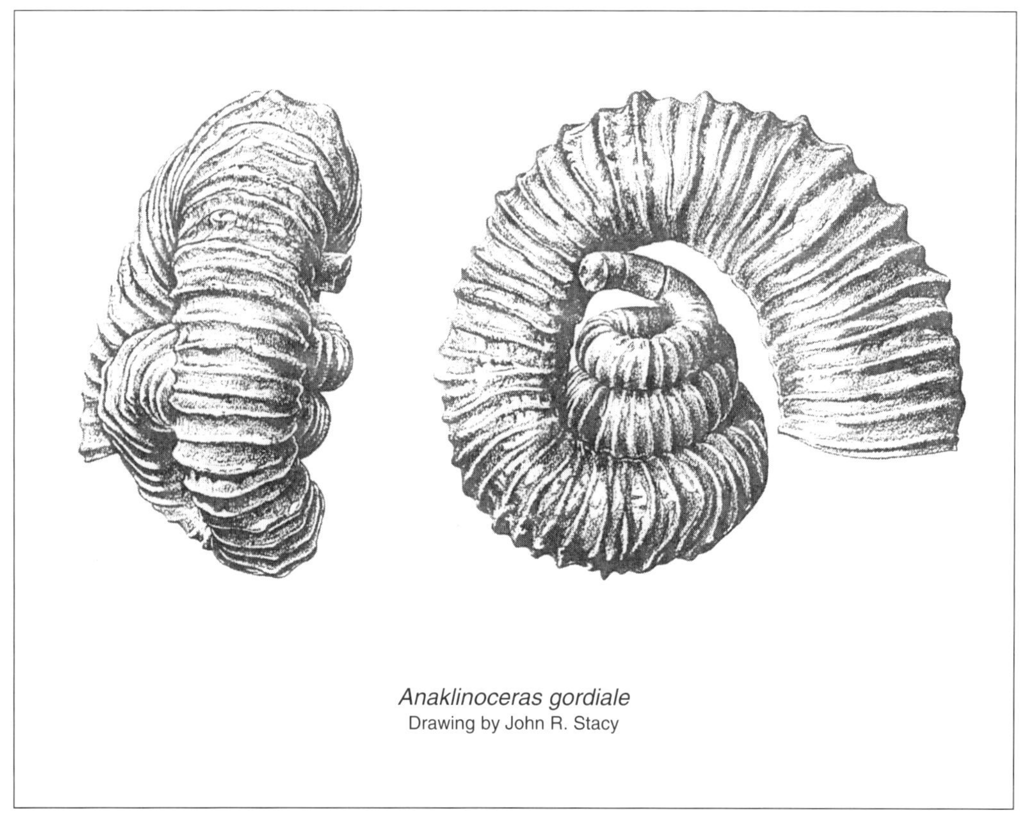

Anaklinoceras gordiale
Drawing by John R. Stacy

ORDER AMMONOIDEA

mmonoidea is an extinct order of cephalopods that flourished in the later Paleozoic and throughout the Mesozoic Epoch, but became extinct at the end of the Cretaceous Period at the same time as the dinosaurs. They are related to the extant and extinct forms of squid, octopi, and nautiloids. Their soft parts include a large head, arms or tentacles, and internal organs. They also had a jaw mechanism that includes the aptychus, which is occasionally preserved. Ammonite shell shapes may be tightly or loosely coiled in a plane, tightly or loosely helically coiled, slightly curved, irregularly curved, or straight.

The Ammonoidea are characterized by an external, multichambered shell with septa that are connected by a tube structure called a siphuncle. The septa, at their intersection with the external shell, form complex, angular suture patterns. The septa and external shell are composed of alternating aragonite and conchiolin, nacreous shell. This layered shell structure gave the thin shell of the ammonite part of its strength to withstand crushing under extreme hydrostatic pressure. Septa created chambers in the phragmocone that enabled these cephalopods to maintain neutral bouyancy. Arrangement of the buoyant hard shell and soft tissues enabled the ammonite shell and visceral mass to maintain a particular orientation in the water (center of buoyancy and center of mass).

All ammonites began life as an embryonic shell stage called an ammonitella, which had a diameter of approximately 1 mm and consisted of a coiled protoconch, one or more septa, and a body chamber. These tiny animals were then able to disperse throughout the seaway, along with other plankton, simply by movement of the water currents until they became large enough to colonize an area of the sea.

The abundance and diversity of the Ammonoidea from the Pierre Shale has provided collectors and scientists with an extensive ammonite fauna that is not yet completely described. To date, there are nine known ammonite families, 28 described genera, and almost 100 described species from this single geologic formation and its marine equivalents in the Western Interior.

The abundance and variety of heteromorph ammonites makes this ancient seaway unique in the world. The excellent preservation of many of the ammonites and other ancient sea life in the Pierre Shale is rarely equaled but never surpassed, if the geographic extent of the formation is considered.

The living ammonitella
[Dashed line indicates phragmocone position]
Illustrations by D.S. Norton

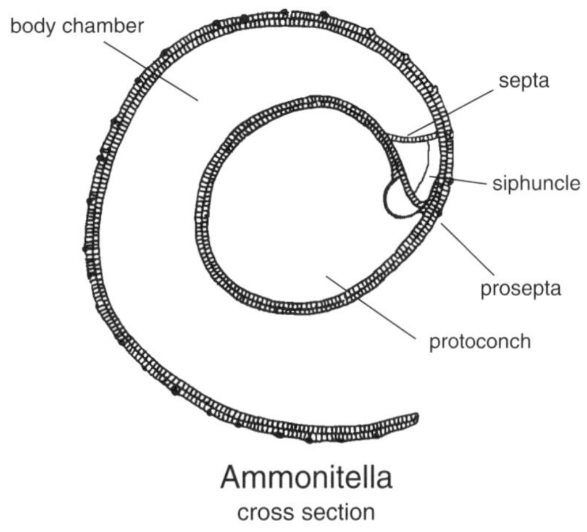

Ammonitella
cross section

The Baculitidae are characterized by an ammonitella followed by a straight to slightly curved shaft. Whorl section can be circular, ovate, elliptical, trigonal, or pear shaped with a flat venter. Ribs consist of undulations or slight swellings on the flank and sometimes prominent ridges on the venter. The degree of taper can vary from low (nearly parallel venter and dorsum) to high (almost triangular). Size ranges from small to quite large (120 cm in length and greater). Eight genera of *Baculites* exist throughout the world, but only two genera have been reported from the Pierre Shale. Baculitidae have a worldwide distribution from the Upper Albian through the Maastrichtian Stage.

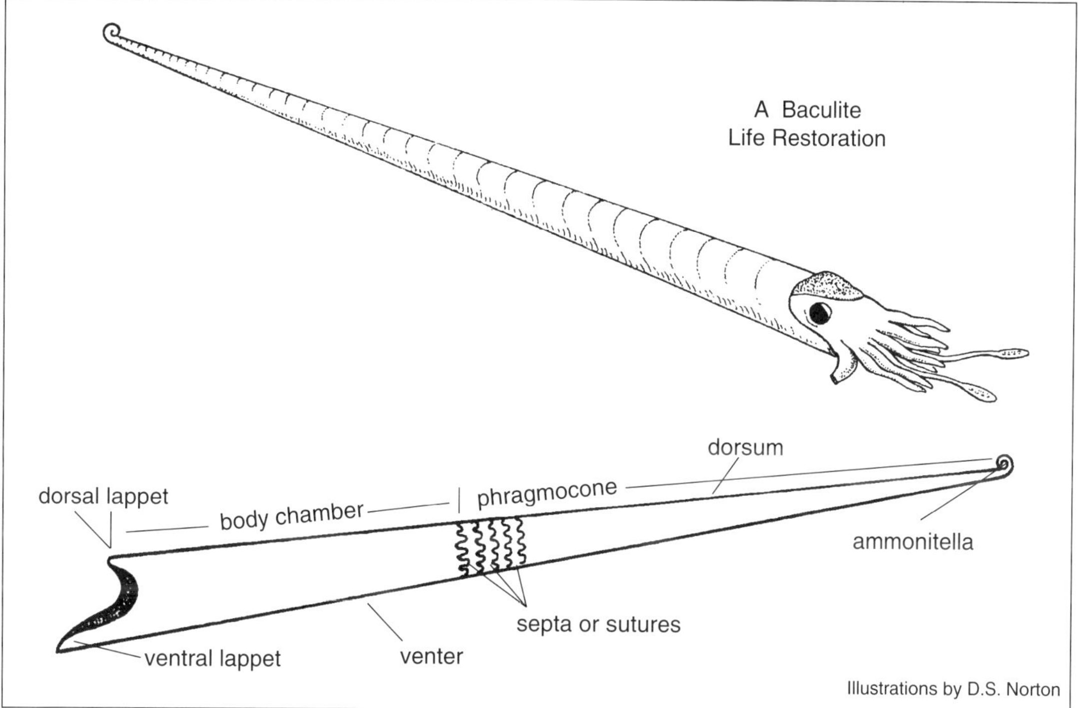

Genus *Baculites* Lamarck, 1799
baculus = a staff + *ites* = a stone

This genus is comprised of a straight to slightly curved shell except for the ammonitella, which has one or two minute, planispiral coils. The cross section can be circular, ovate, elliptical, compressed elliptical, or triangular. The adult aperture has a short, dorsal lappet and a longer, ventral lappet that is straight to slightly curved dorsally. The venter commonly has ribs, and the flanks may be decorated by broad ribs, rounded to concentric nodes or broad undulations. The degree of taper generally varies from $0°$ to $8°$, but most species exhibit a degree of taper of about $2°$. The suture ranges from simple in the early forms to very complex at the time of *Baculites rugosus*. Individual species of *Baculites* are very important index fossils in the Pierre Shale, comprising 18 of the 26 ammonite zones. The following baculite descriptions are listed in chronological order from the oldest to the youngest Range Zone. Also included are the other common species of *Baculites*. For identification purposes, a moderate- or average-size baculite has a diameter of approximately 2.5 to 5 cm in a mature body chamber.

Baculites aquilaensis Reeside, 1927

Baculites aquilaensis is distinguished by a compressed ovate cross section, medium size (generally smaller than *Baculites obtusus*), strong and short arcuate ribbing that occupies only about half of the flank and its early appearance of ribs. The prominent arcuate ribs occur about 2.5 per flank width. The venter is slightly ribbed and compressed, while the dorsum is smooth and rounded. This species occurs with *Scaphites hippocrepis* in the basal Pierre Shale and its equivalents (the Steele Shale, Mancos Shale, the Eagle Sandstone, and the Telegraph Creek Formation) in Montana, Wyoming, South Dakota, Colorado, New Mexico, and Utah.

Baculites haresi Reeside, 1927

Baculites haresi has a very elliptical, well-rounded cross section. In large, mature specimens the shell is as round on the venter as in the dorsum. Younger specimens tend to have a slightly compressed venter. The shell is smooth, although the ventral edge can be slightly ribbed. This species occurs with *Scaphites hippocrepis* from the Eagle Sandstone, the Telegraph Creek Formation, the Steele Shale, the upper part of the Mancos Shale and the Pierre Shale of Montana, Wyoming, South Dakota, Utah, and New Mexico.

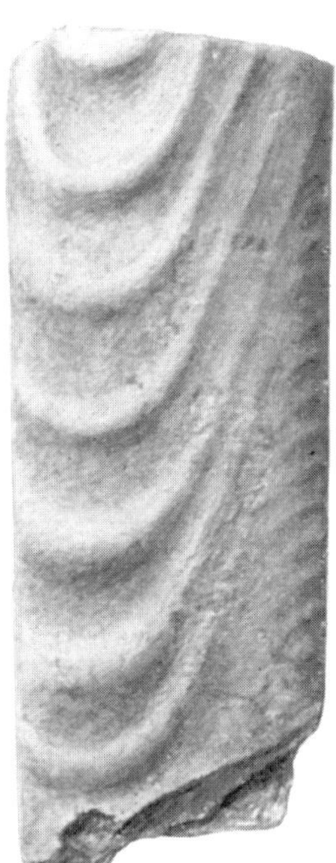

Baculites aquilaensis
Side view
USNMH 73298
6.8 cm long

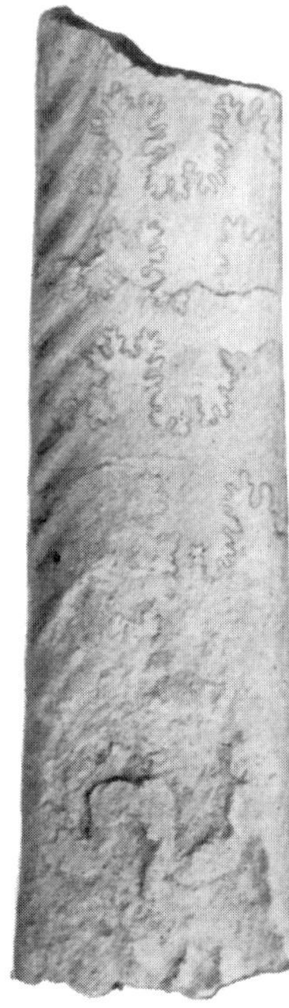

Baculites haresi
Side view
USNMH 73298
7.6 cm long

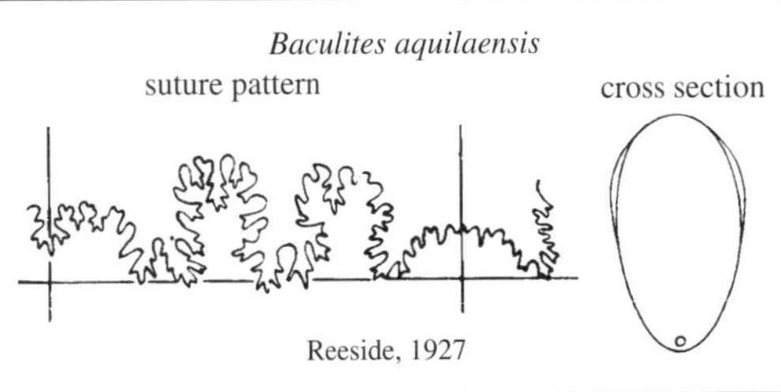

Baculites aquilaensis

suture pattern
cross section

Reeside, 1927

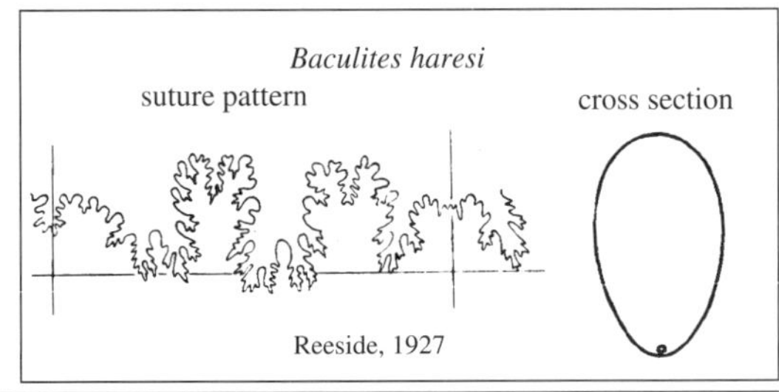

Baculites haresi

suture pattern
cross section

Reeside, 1927

Baculites **sp.** (smooth)

Baculites sp. (smooth)
USGS D 3550 6.3 cm long

This species is the first (oldest) of the baculites to be used as a Range Zone fossil. Specimens have an ovate cross section and a low degree of taper. They are generally quite smooth on the flanks, hence, the informal name. The venter can reveal some minor ribbing. *Baculites* sp. (smooth) reaches 3 to 5 cm in diameter and 15 to 30 cm in length. This baculite and all of the successive species up to *Baculites gregoryensis* have similar, fairly simple suture patterns. They have been reported from the Lower Pierre Shale and equivalent-age rocks in Colorado, Montana, South Dakota, and Wyoming.

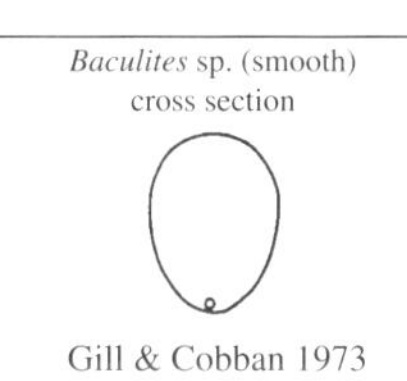

Baculites sp. (smooth)
cross section

Gill & Cobban 1973

Enhanced View

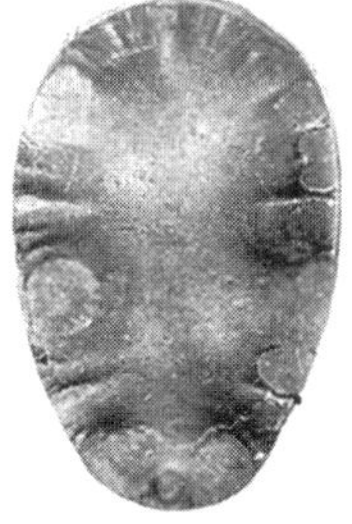

Baculites sp. (smooth)
USGS D 3550 cross section

Baculites **sp.** (weak flank ribs)

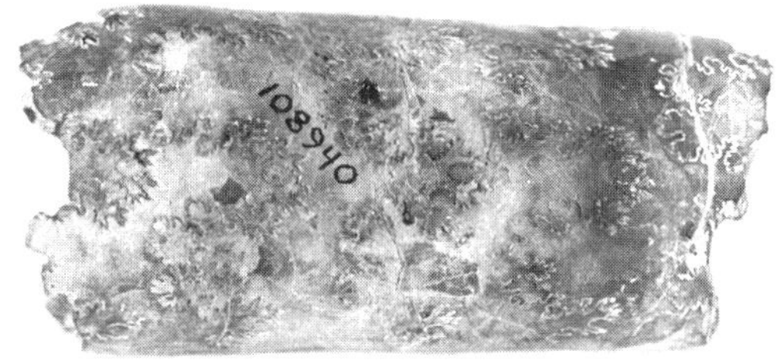

Baculites sp. (weak flank ribs)
USNM 23639 18 cm long

Directly above the *Baculites* sp. (smooth), in the Gammon Ferruginous Member of the Pierre Shale and equivalent formations, is the unnamed baculite, *Baculites* sp. (weak flank ribs). It is characterized by low arcuate ribs on the flank and a smooth to well-ribbed venter. *Baculites* sp. (weak flank ribs) is a moderate-sized baculite similar to its predecessor in size, shape, taper, and suture pattern. They are reported from the Pierre Shale and equivalent-age formations in Colorado, Wyoming, South Dakota, and Montana.

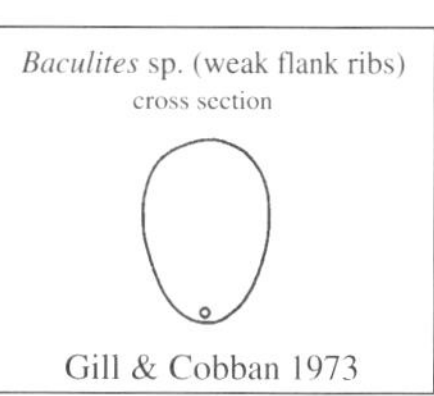

Baculites sp. (weak flank ribs)
cross section

Gill & Cobban 1973

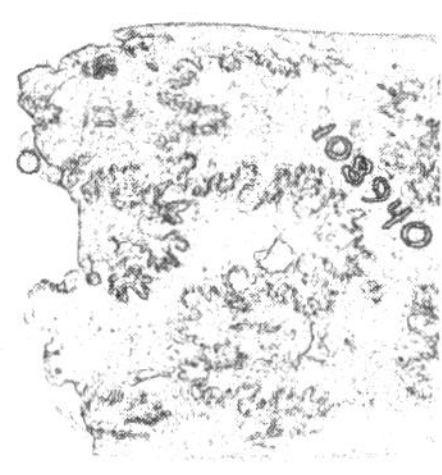

Enhanced View

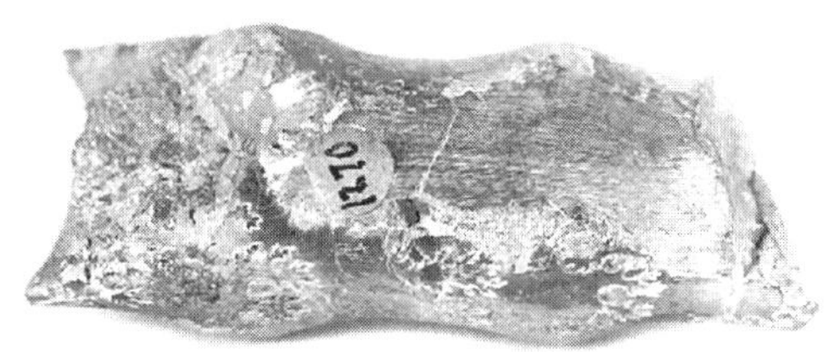

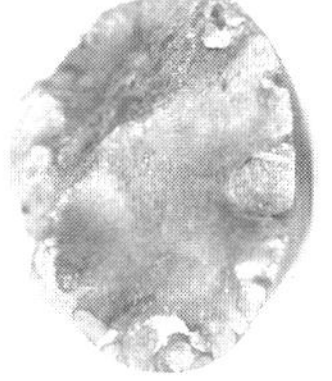

Baculites sp.
(weak flank ribs)

Baculites obtusus Meek, 1876

Baculites obtusus is a small- to moderate-sized conch that has very little taper, an ovate cross section, smooth to well-ribbed venter, and nodelike flank ribs or blunt nodes spaced about every two ribs for the shell diameter. Its suture is of average complexity and has rectilinear lobes and saddles. This species is found in the Sharon Springs Member of the Pierre Shale in South Dakota, Wyoming, Colorado, and New Mexico, as well as in the Claggett Shale in Montana.

Baculites mclearni Landes, 1940

Baculites mclearni has a moderate ovate cross section. The species is characterized by widely spaced, broad flank ribs, and a smooth to weakly ribbed venter. *Baculites mclearni* combines features of the earlier *Baculites obtusus* and the later *Baculites asperiformis* in its ribs and low degree of taper. The suture pattern is moderately complex. Specimens are found in the upper part of the Sharon Springs Member and lower part of the Mitten Member in Wyoming, South Dakota, and Colorado, and in the Claggett Shale in Montana.

Baculites obtusus
USGS D2624
cross section

Baculites obtusus
Ventral view
USGS D2624
9.5 cm long

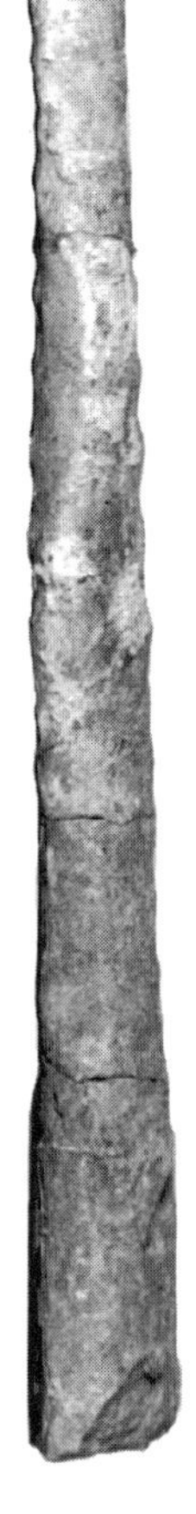

Baculites mclearni
USGS D6372
cross section

Baculites mclearni
USGS D6372
35 cm long

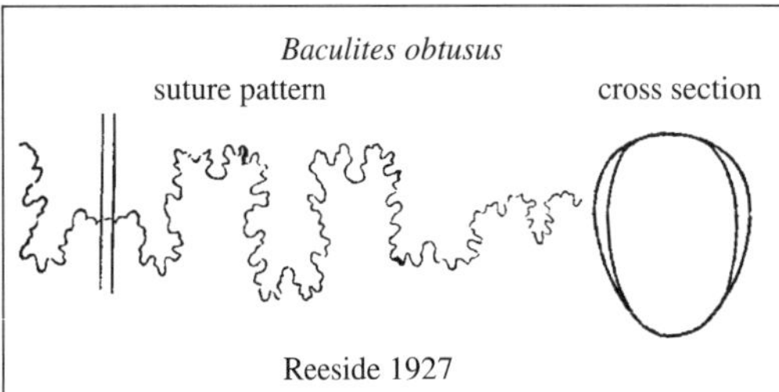

Baculites obtusus
suture pattern · cross section

Reeside 1927

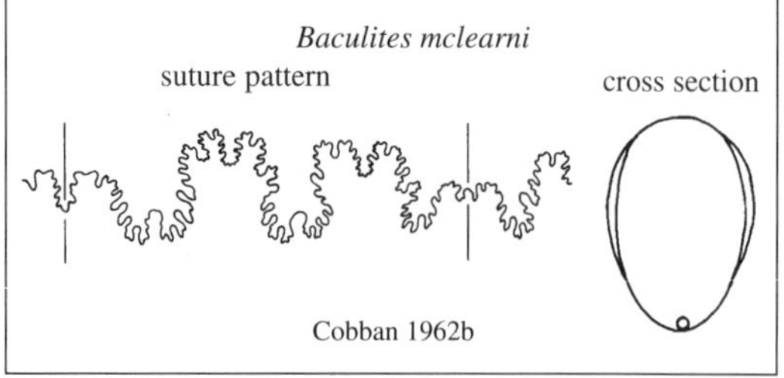

Baculites mclearni
suture pattern · cross section

Cobban 1962b

Baculites asperiformis Meek, 1876

Baculites asperiformis also has a small degree of taper, ovate cross section, smooth to well-ribbed venter, and rounded flank nodes spaced about one for every distance of the shell diameter. Meek described it as having "strong, oblique, nearly straight ridges or undulations, extending entirely across the sides." The suture is of average complexity. This species has been found in the upper part of the Sharon Springs Member in Kansas, Colorado, Wyoming, and South Dakota, the Claggett Shale of Montana, and in the Lewis Shale in New Mexico.

Right:
Baculites asperiformis
USGS D8224
27.5 cm long

Baculites asperiformis
USGS D8224
cross section

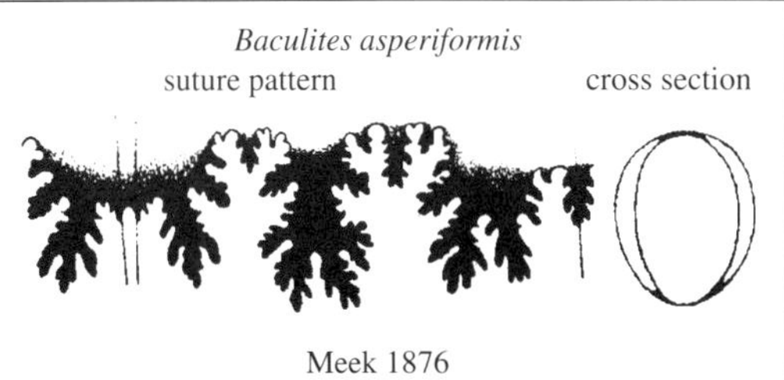

Baculites asperiformis
suture pattern   cross section

Meek 1876

Baculites **sp.** (smooth species)

Baculites sp. (smooth species) is an unnamed species found just above the top of the Sharon Springs Member in eastern Wyoming. It has a low degree of taper, a subelliptical cross section, and is smooth on the flanks and venter. It is moderate to large in size with a suture of average complexity. A similar species of about the same age from Russia was described as *Baculites cobbani* Khakimov in 1976 and was supposed to be the name for *Baculites* sp. (smooth species). However, there is a noticeable difference in the constrictions on the middle lobe of the suture of the Russian form, which makes this name, *Baculites cobbani*, doubtful for the species in the Western Interior.

Baculites sp.
(smooth species)
USGS D2140
cross section

Baculites sp.
(smooth species)
USGS D2140
14.3 cm long

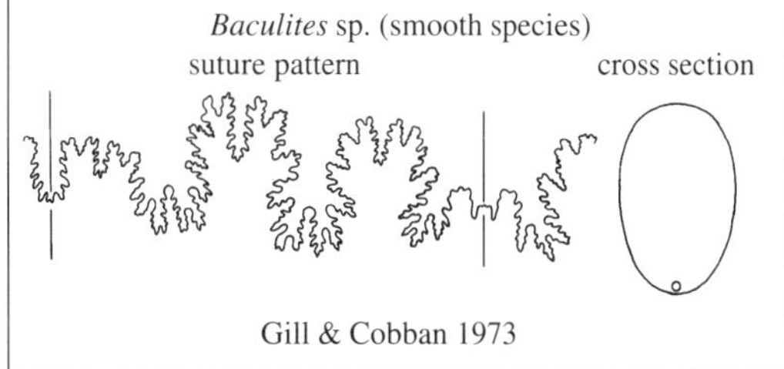

Baculites sp. (smooth species)
suture pattern   cross section

Gill & Cobban 1973

Baculites perplexus Cobban, 1962b

Baculites perplexus has a gentle to low taper and a corrugated venter with smooth to weakly sculptured flanks. The prominent ribs on the venter average about four ribs for the shell diameter. The suture is of average complexity. The species is larger than average size for the genus. Specimens have been found in the Mitten Member of the Pierre Shale, around the Black Hills, in the Steele Shale in central Wyoming, in the Claggett Member of the Cody Shale in south central Montana, in the upper part of the Mancos Shale in northwest Colorado, in the Pierre and Lewis Shales in New Mexico, and in equivalent strata in eastern Utah.

In some areas there is an upper and lower Range Zone of *Baculites perplexus* separated by *Baculites gilberti*.

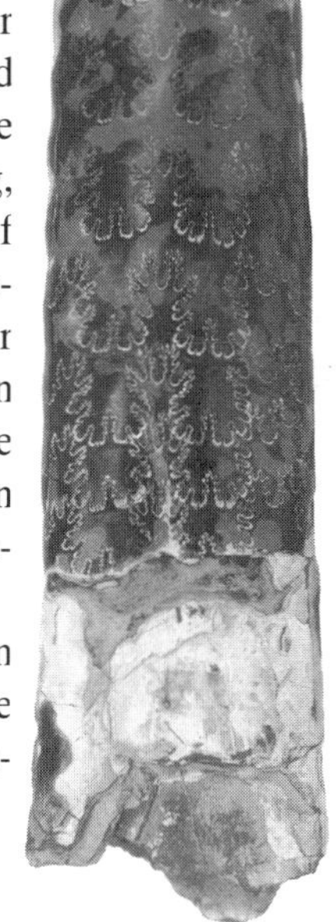

Baculites perplexus
Above: BHI 4763 9.5 cm long
Left: BHI 4762 5.5 cm long

Baculites perplexus
USGS D6251
cross section

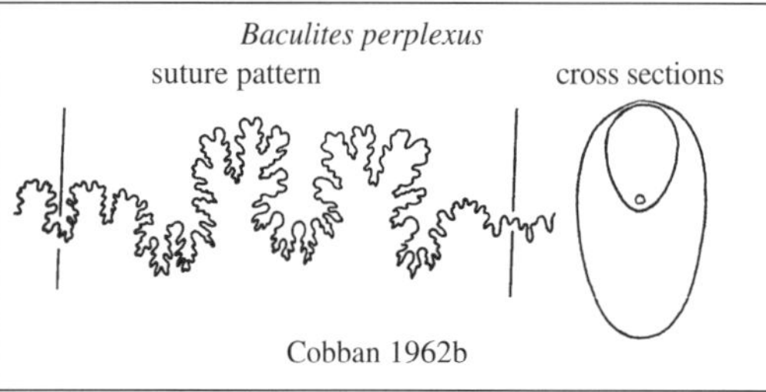

Baculites perplexus
suture pattern   cross sections

Cobban 1962b

Baculites gilberti Cobban, 1962b

Baculites gilberti is found between an early and late form of *Baculites perplexus*. This moderate-sized baculite has a ribbed venter

Baculites gilberti
USGS D7804 9 cm long

with average spacing of six ribs per shell diameter. Undulations on the flanks are spaced at two per shell diameter. It has a subovate cross section in juveniles and an elliptical one in adults. The suture is similar to *Baculites perplexus*, but the ribbing on the venter is considerably weaker and more closely spaced. It occurs in the upper part of the Mancos Shale in northwest Colorado, in the "rusty zone" of the Pierre Shale in Colorado and Wyoming, and in the Mitten and Red Bird Members of the Pierre Shale around the Black Hills.

Baculites gilberti
USGS D7804
cross section

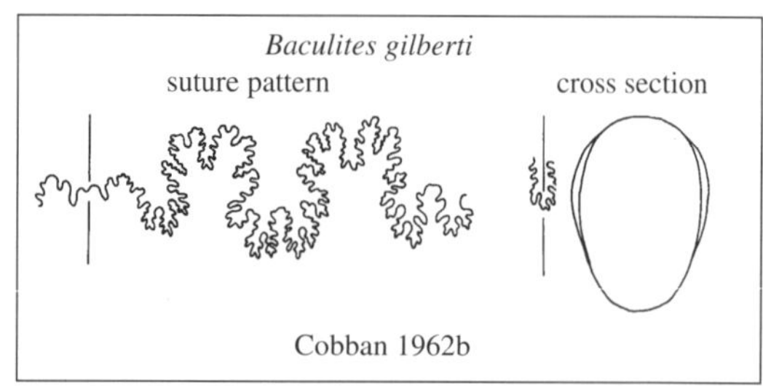

Baculites gilberti
suture pattern   cross section

Cobban 1962b

Baculites gregoryensis Cobban, 1951a

Baculites gregoryensis is a rapidly tapering, moderately sized ammonite. It has an oval cross section that is quite compressed on the ventral edge. The flanks are smooth, and the venter can have slight ribbing or corrugations, especially on larger specimens. The suture is fairly com-

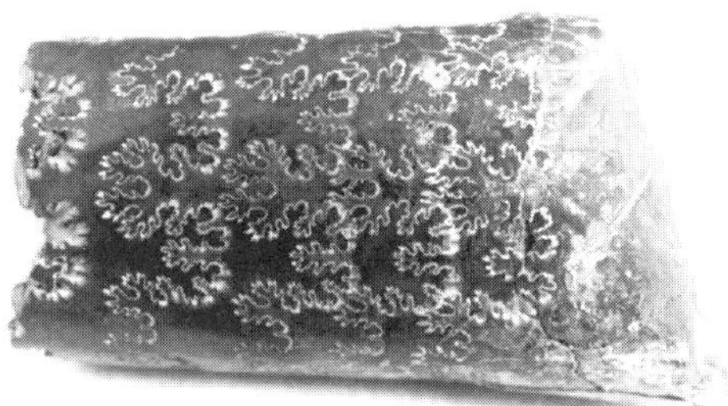

Baculites gregoryensis
BHI 4163 6.5 cm long

plex and characterized by the lateral lobe being constricted just above its major branches. *Baculites gregoryensis* is found in the Gregory Member of the Pierre Shale along the Missouri River in South Dakota, above the Mitten Member on the northern and southern flanks of the Black Hills, and in central Montana.

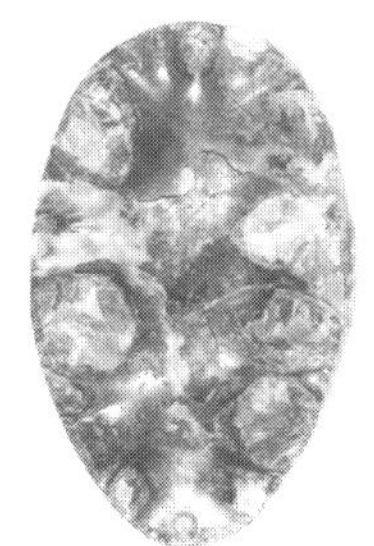

Baculites gregoryensis
BHI 4163
cross section

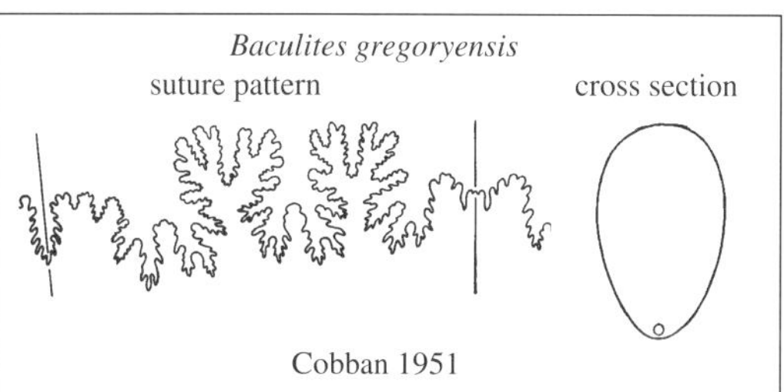

Baculites gregoryensis
suture pattern   cross section

Cobban 1951

Baculites reduncus Cobban, 1977

Baculites reduncus is a rapidly tapering, moderately sized, curved ammonite. Its cross section is ovate, yet slightly compressed on the ventral edge. The flanks are ornamented with broad arcuate ribs or undulations. The venter is smooth to slightly ribbed. The suture is similar to that of *Baculites gregoryensis.* The species has been found in the lower part of the Rock River Formation in Wyoming, in southwestern and central South Dakota, and in equivalent-age rocks in Colorado.

Baculites reduncus
BHI 4760
12.5 cm long

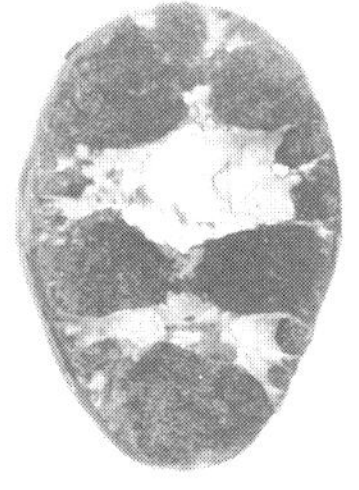

Baculites reduncus
USGS D1392
cross section

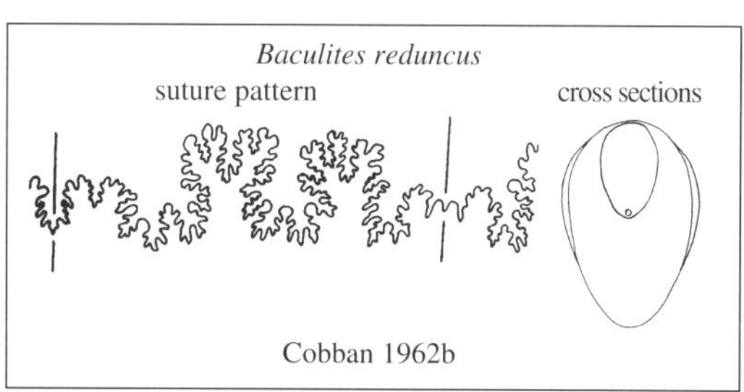

Baculites reduncus
suture pattern   cross sections

Cobban 1962b

Baculites scotti Cobban, 1958a

Baculites scotti is gently tapering and of average size. The cross section is ovate yet slightly compressed on the ventral edge. The flanks are generally smooth to slightly ribbed. The suture of *Baculites scotti* can be more complex than that of *Baculites gregoryensis*. *Baculites scotti* has been found in the Pierre Shale in New Mexico, Utah, Colorado, Wyoming, and South Dakota, in the Mancos Shale in western Colorado; and in a marine tongue in the Judith River Formation in east central Montana.

Baculites scotti
BHI 4168
16 cm long

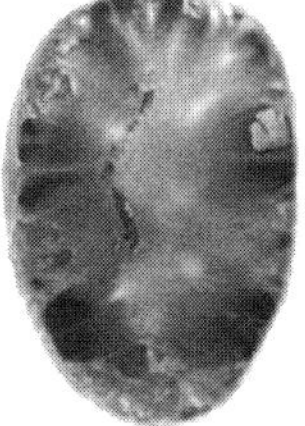

Baculites scotti
BHI 4168
cross section

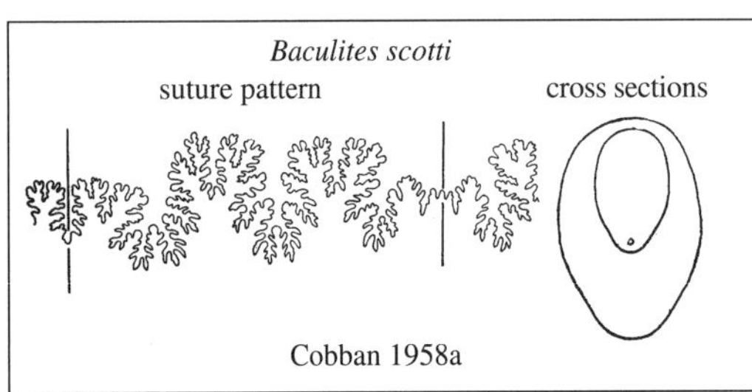

Baculites scotti
suture pattern   cross sections

Cobban 1958a

Baculites sp. (new species)

Occurring with *Baculites scotti* is a baculite form that is ovate in cross section, has a smooth venter, and has broad ribs or arcuate swellings on the flanks. The broad ribs become almost like a large cone-shaped tubercle in the smaller specimens and appear nodelike. These ribs occur about once for every flank diameter. The suture pattern is like that of its contemporary, *Baculites scotti*. This species has been described, but has not yet been published. This unpublished species occurs throughout the Western Interior along with *Baculites scotti*.

Baculites sp. (new species)
BHI 4671
2 cm by 5 cm long

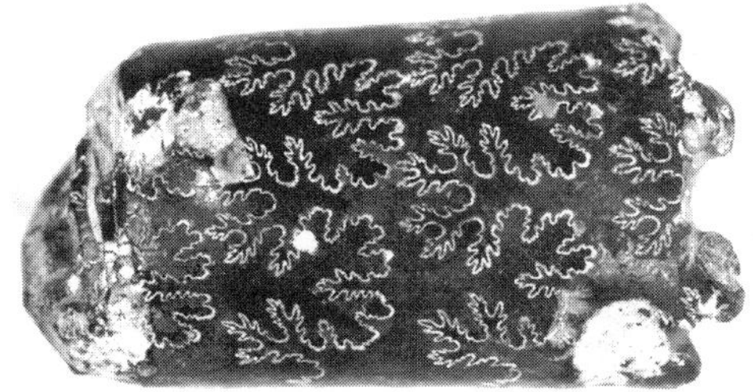

Baculites sp. (new species)
BHI 4671
3 cm by 5 cm long

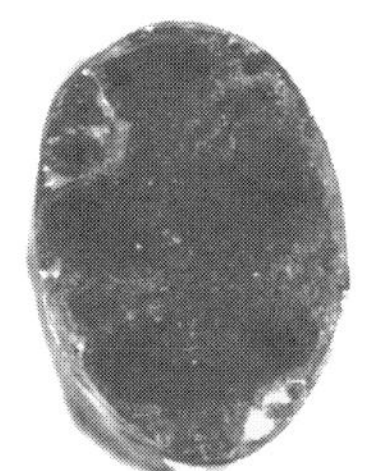

Baculites sp. (new species)
BHI specimen
cross section

Baculites pseudovatus Elias, 1933

Baculites pseudovatus is the predominant *Baculites* found throughout the Range Zone of *Didymoceras nebrascense*. It is a moderate to large-sized species that has a gentle taper, a stout ovate cross section, smooth flanks on most specimens, and smooth to weakly ribbed venters. Larger specimens have undulations on the flanks. The complex suture has the terminal branches of the lateral lobe constricted at their base. Specimens have been found in Montana, South Dakota, Wyoming, Nebraska, Kansas, Colorado, and New Mexico.

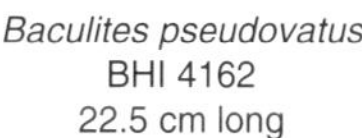

Baculites pseudovatus
BHI 4162
22.5 cm long

Baculites pseudovatus
BHI 4162
cross section

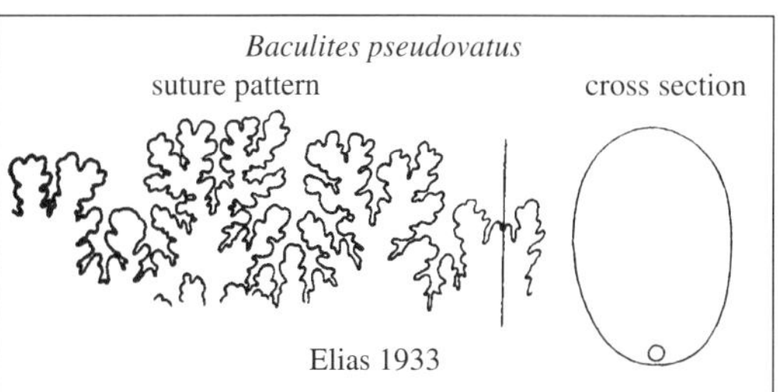

Baculites pseudovatus
suture pattern

cross section

Elias 1933

Baculites crickmayi Williams, 1930

Baculites crickmayi has an ovate cross section, a rapidly tapering conch in juveniles, and a slowly tapering shell in mature adults. On larger specimens, broad undulations ornament the flank. The venter may be smooth or weakly to strongly ribbed. The species is most often found with *Didymoceras stevensoni* in New Mexico, Colorado, Wyoming, Montana, and Canada. The suture pattern of *Baculites crickmayi* is complex and similar to that of *Baculites pseudovatus*.

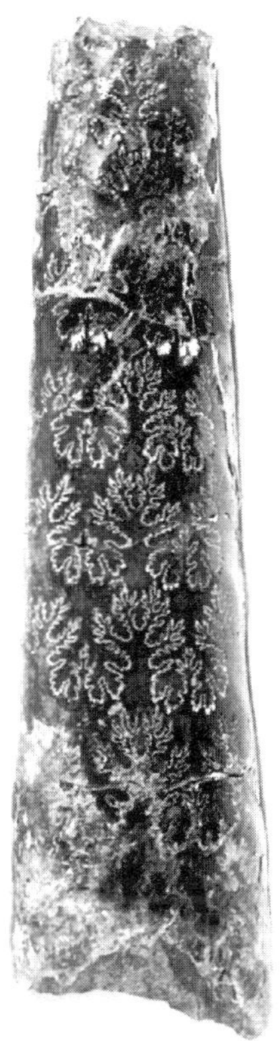

Baculites crickmayi
BHI 4165
19 cm long

Baculites crickmayi
BHI 4165
cross section

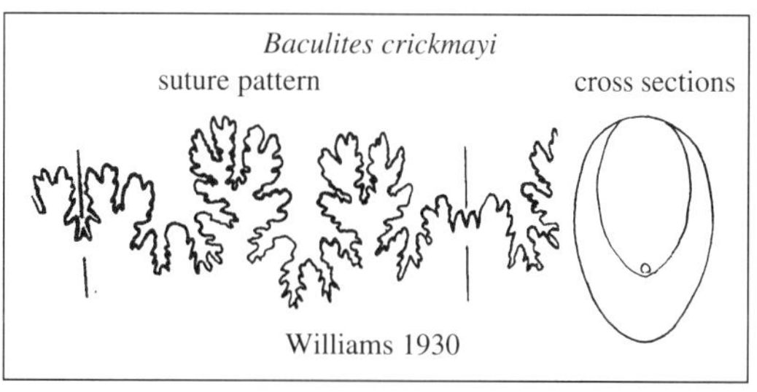

Baculites crickmayi
suture pattern

cross sections

Williams 1930

Baculites rugosus Cobban, 1962a

Baculites rugosus ranges from a large, laterally compressed, rapidly tapering form to a smaller, stouter, more ovate form. The larger, more common variety has a flattened ovate cross section, strong ventral ribbing, and a very complex suture pattern. This was the largest of the Baculite species during Campanian time. The smaller, immature specimens are more ovate in cross section, but as they grow, they flatten on the flank, and the ventral ribbing becomes very prominent, with ribs spaced three to four for the shell diameter. The species is associated with *Exiteloceras jenneyi*. *Baculites rugosus* is widely distributed from New Mexico to Montana.

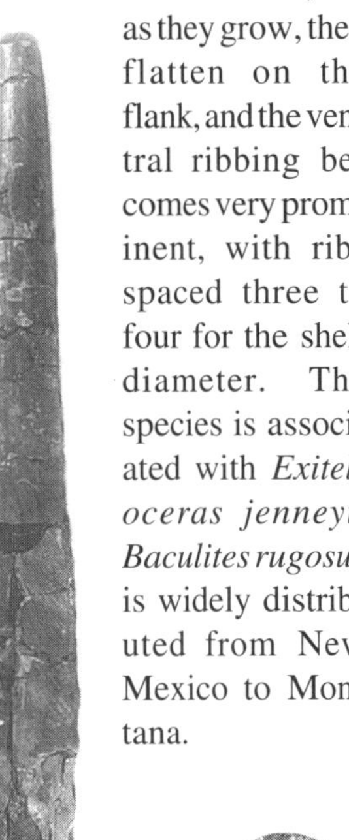

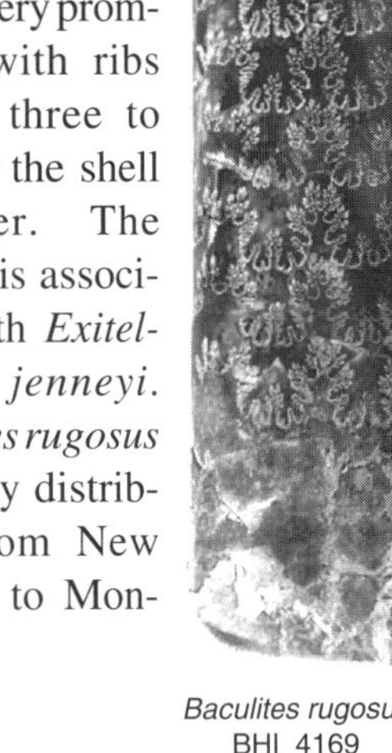

Baculites rugosus
BHI 4170
43.5 cm long

Baculites rugosus
BHI 4169
22.5 cm long

Left:
cross section

Baculites corrugatus Elias, 1933

Baculites corrugatus is found in the *Didymoceras cheyennense* Range Zone. It has a compressed to elliptical cross section similar to, yet more ovate than, the later *Baculites compressus*. The venter has pronounced corrugations that are spaced about every five to seven ribs for the shell diameter. The shell tapers slowly to form a long, nearly straight shaft. The suture pattern is similar to *Baculites compressus,* but has an inclined second lobe. *Baculites corrugatus* is found within the zone of *Didymoceras cheyennense* from New Mexico, Kansas, Colorado, South Dakota, Wyoming, and Montana.

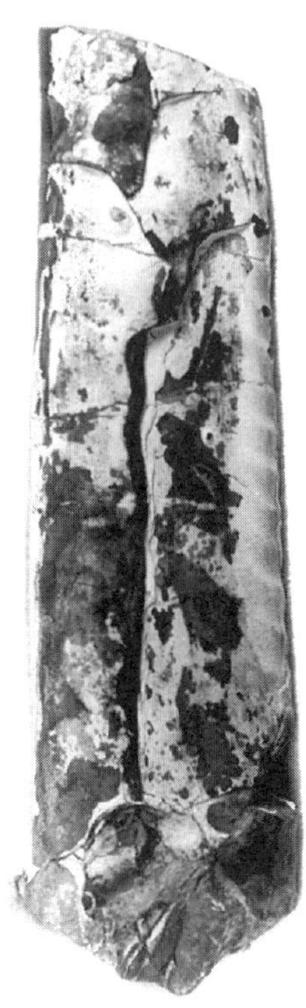

Baculites corrugatus
BHI 4161
14 cm long

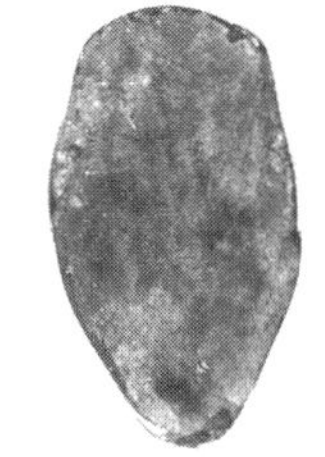

Baculites corrugatus
BHI 4161
cross section

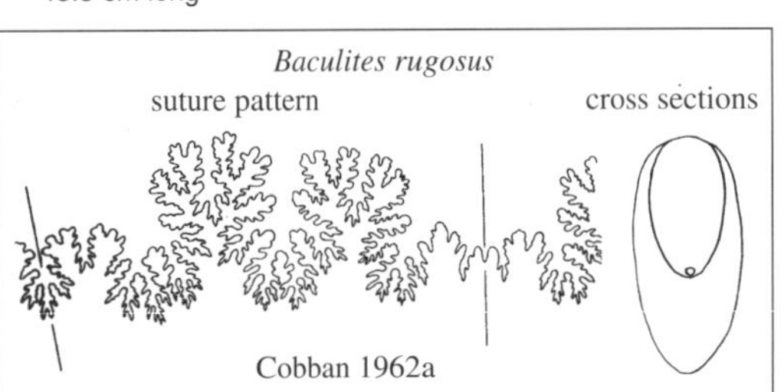

Baculites rugosus
suture pattern
cross sections
Cobban 1962a

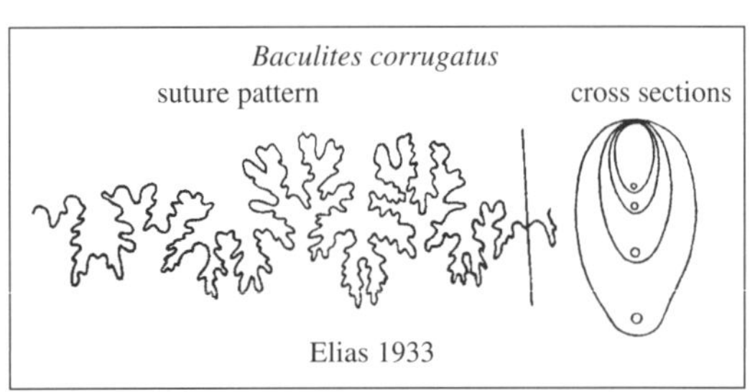

Baculites corrugatus
suture pattern
cross sections
Elias 1933

Baculites compressus Say, 1820

Baculites compressus is the most abundant ammonite in the middle Pierre Shale within South Dakota. It is easy to identify because of its very compressed (hence the name), elliptical cross section and smooth to weakly ribbed venter. Large specimens have slight undulations on the flank. The species tapers rapidly and can curve somewhat in juveniles. It is almost straight in mature adults. *Baculites compressus* has a very complex suture pattern. Specimens can attain a large size and, within their zone, are the most prolific ammonite, if not the most abundant mollusk that is found. The species ranges from Colorado into Canada.

Baculites compressus
BHI 4159
17 cm long

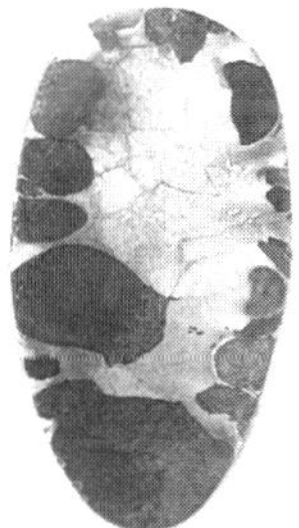

Baculites compressus
BHI 4159
cross section

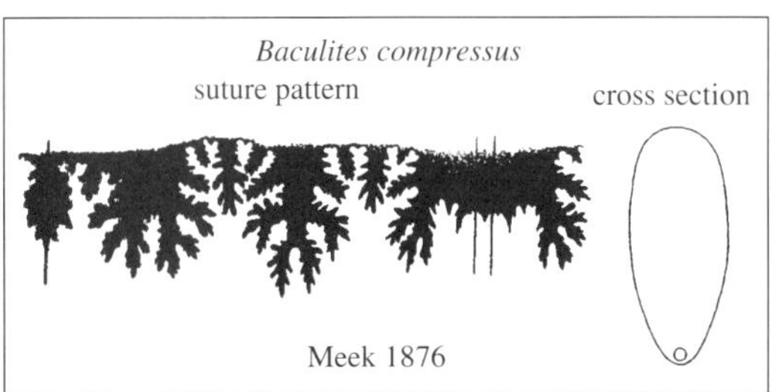

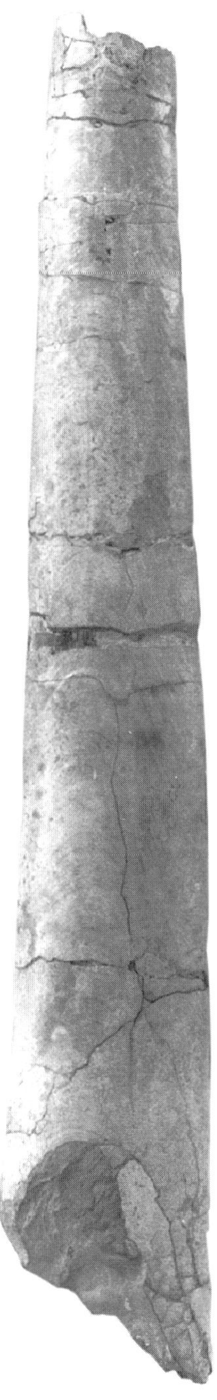

Baculites compressus
BHI 4160
40 cm long

Baculites compressus **var.** *robinsoni* Cobban, 1962a

Baculites compressus var. *robinsoni* was originally described as *Baculites* var. *ornatus* by Robinson, 1945. However, Cobban renamed it after Dr. Robinson because the type *Baculites ornatus* was described by d'Orbigny, 1947, from South America. This variety has a slightly greater degree of taper than that of *Baculites compressus*. The cross section is compressed to trigonal in the immature specimen to nearly elliptical in mature specimens, but not as trigonal as *Baculites cuneatus* nor as compressed as *Baculites compressus*. The flanks are ornamented with broad arcuate undulations similar to those of *Baculites cuneatus,* but occur at an earlier stage of growth. The suture pattern is less complex than that of either *Baculites compressus* or *Baculites cuneatus*.

Baculites compressus var. *robinsoni* occurs from the *Baculites compressus* and *Baculites cuneatus* Zones of the Bearpaw Shale in Saskatchewan, south-western Alberta, and central Montana. The species has also been reported in the Pierre Shale of South Dakota. This is often the most abundant baculite found within the middle section of the Bearpaw Shale.

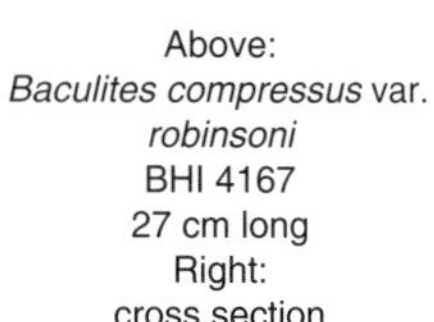

Above:
Baculites compressus var.
robinsoni
BHI 4167
27 cm long
Right:
cross section

Above:
Baculites compressus var. *robinsoni*
with *Baculites compressus*
BHI 4756
41.5 cm total length

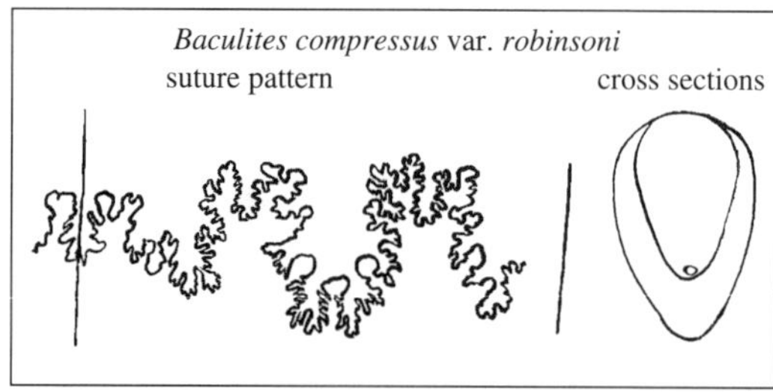

Baculites compressus var. *robinsoni*
suture pattern   cross sections

Baculites undatus Stephenson, 1941

Baculites undatus has a stout elliptical to round cross section. The flanks have broad arcuate ribs or undulations with an average spacing of 1.4 per shell diameter. The venter is broad and also displays broad to fine ribbing. The suture pattern is much less complex than that of its contemporaries, and is nearly identical to that of *Baculites baculus. Baculites undatus* is small in size for the genus, the diameter of an adult specimen measuring one to two inches.

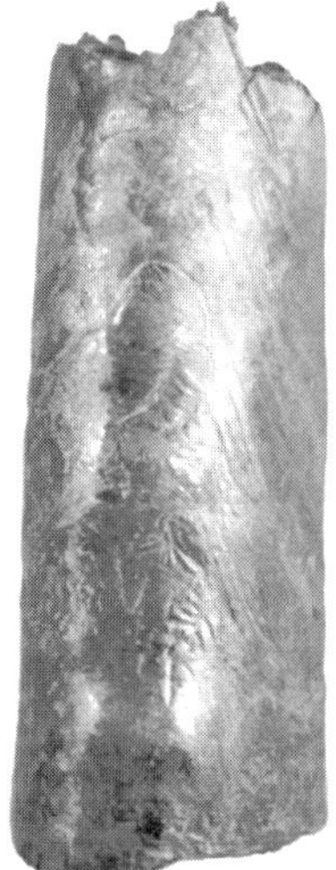

This species was originally described from the Nacatoch Sand of northeastern Texas and has also been found in Delaware, New Jersey, Tennessee, and Mississippi. Cobban, 1973, reported the occurrence of this Gulf Coast baculite within the Western Interior in New Mexico and north into northern Colorado from the *Didymoceras cheyennense* Zone through the *Baculites reesidei* Zone. More recently, *Baculites undatus* has been reported from the *Baculites compressus* and *Baculites cuneatus* Zones in Meade and Pennington Counties in South Dakota.

Baculites undatus
Above: BHI 4034
lateral view
11.5 cm high

Below:
USNM 182427
cross section

Baculites cuneatus Cobban, 1962a

Baculites cuneatus is large for the genus. It is mainly characterized by its cuneiform or triangular cross section and narrow venter. When young, the juveniles are often curved with smooth flanks and a high degree of taper. Upon maturity they become fairly straight with a low degree of taper. Mature specimens have well-developed undulations on the flank and slight to moderate ribs on the venter. The species has a very complex suture pattern like that of *Baculites compressus. Baculites cuneatus* is very common from Colorado into Canada.

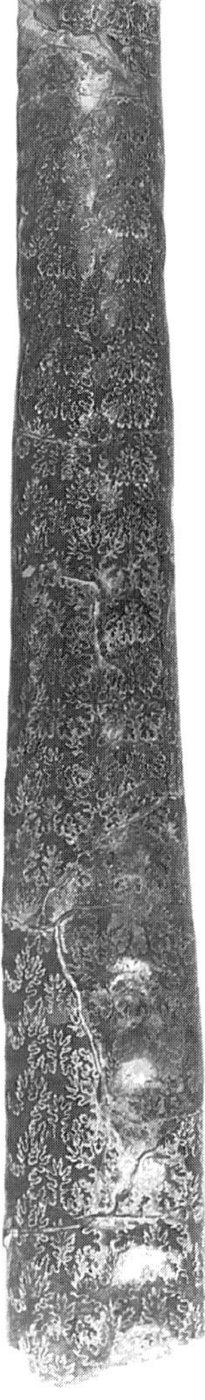

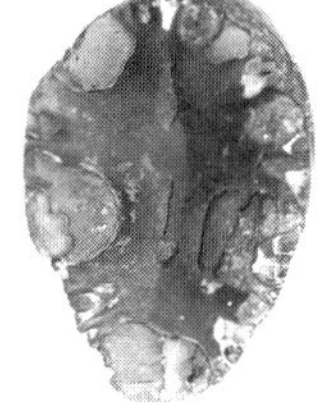

Baculites cuneatus
BHI 4158
cross section

Baculites cuneatus
BHI 4158
49 cm long

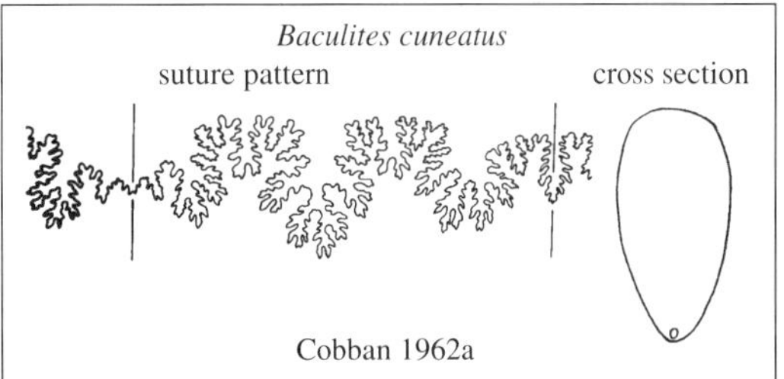

Baculites undatus
suture pattern   cross section

Cobban 1973

Baculites cuneatus
suture pattern   cross section

Cobban 1962a

Baculites reesidei Elias, 1933

Baculites reesidei is very similar to *Baculites compressus* in shape but with a compressed triangular cross section, and a slightly less complex suture pattern. Adults generally have a slight degree of taper, but young juvenile shells can taper rapidly. The ventral edge has larger and wider spaced ribs that form slight corrugations that distinguish it from its predecessor, *Baculites compressus*. *Baculites reesidei* has been recorded from Kansas, Colorado, Wyoming, South Dakota, Montana, Alberta, and Saskatchewan.

Baculites reesidei
BHI 4157
23 cm long

Baculites reesidei
BHI 4157
cross section

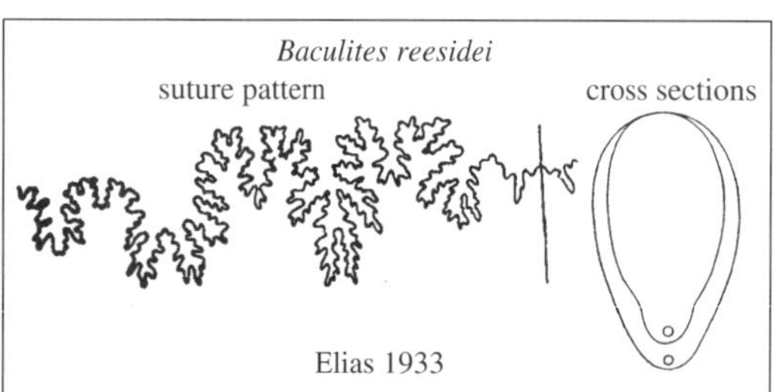

Baculites reesidei
suture pattern   cross sections

Elias 1933

Baculites jenseni Cobban, 1962a

Baculites jenseni is large and has an ovate cross section. It is characterized by its low degree of taper, smooth flanks, and a dorsolateral depression. Young adults have ribs on the ventral edge (about nine per shell diameter), but older more mature specimens seem to be smoother. The suture is similar to that of *Baculites compressus*. Specimens have been recorded from the Pierre Shale of Colorado, South Dakota, and Wyoming, the Bearpaw Shale of Montana and Canada, and the Lewis Shale of Wyoming.

Left:
Baculites
jenseni
BHI 4045
7.5 cm long

Left:
Baculites jenseni
USGS D8018
cross section

Above:
Baculites jenseni
BHI 4757
33 cm long

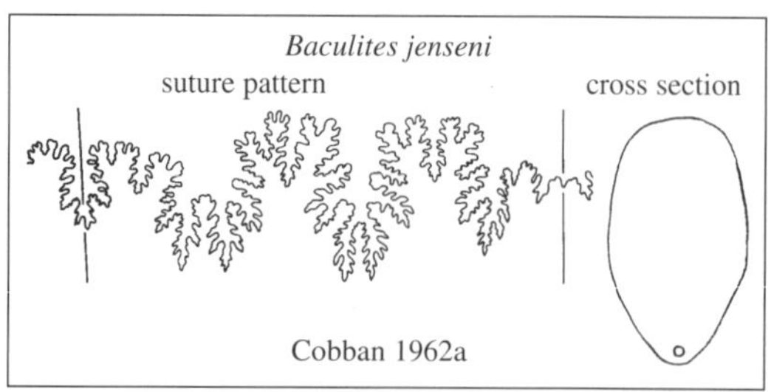

Baculites jenseni
suture pattern   cross section

Cobban 1962a

Baculites eliasi Cobban, 1958a

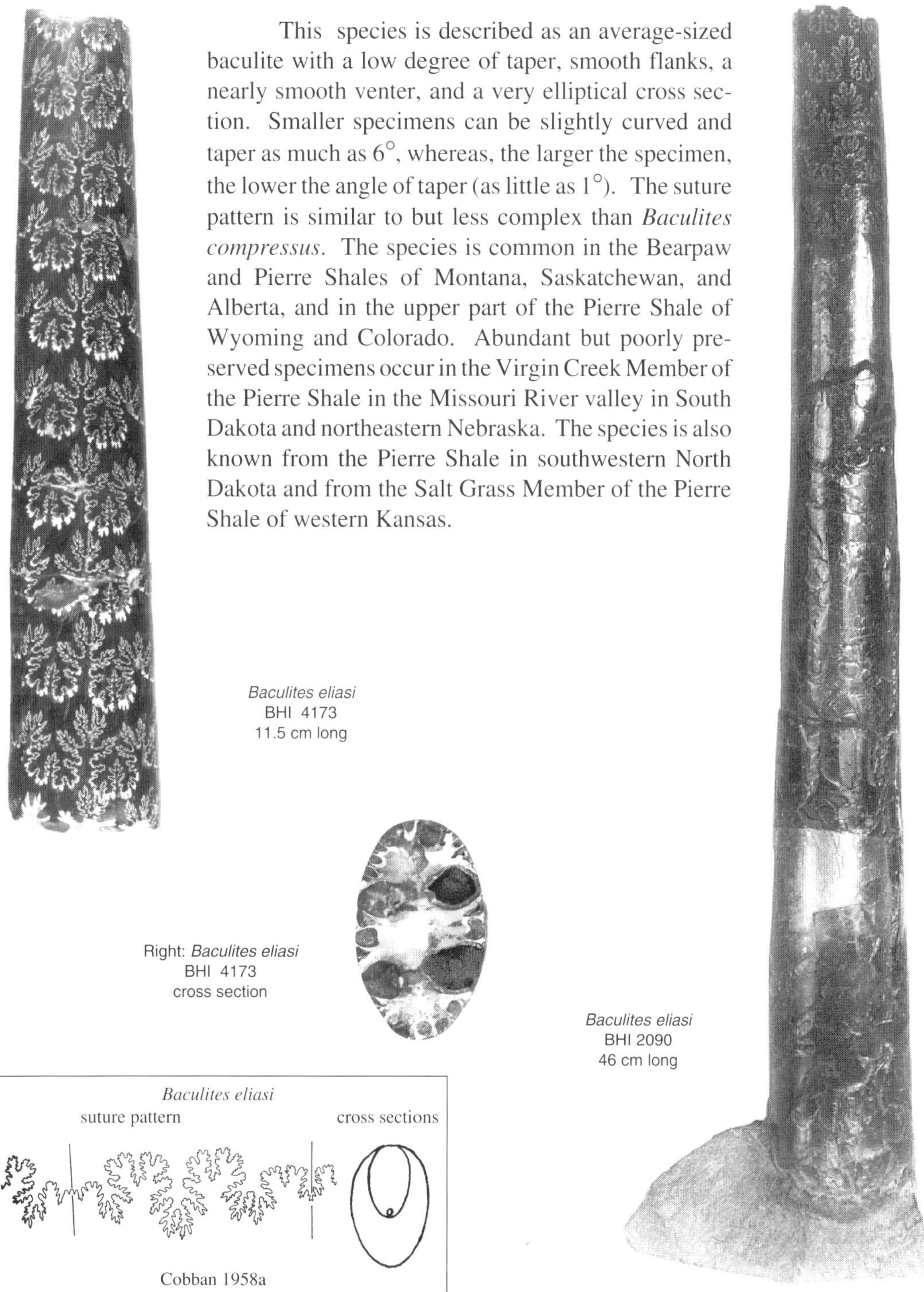

This species is described as an average-sized baculite with a low degree of taper, smooth flanks, a nearly smooth venter, and a very elliptical cross section. Smaller specimens can be slightly curved and taper as much as 6°, whereas, the larger the specimen, the lower the angle of taper (as little as 1°). The suture pattern is similar to but less complex than *Baculites compressus*. The species is common in the Bearpaw and Pierre Shales of Montana, Saskatchewan, and Alberta, and in the upper part of the Pierre Shale of Wyoming and Colorado. Abundant but poorly preserved specimens occur in the Virgin Creek Member of the Pierre Shale in the Missouri River valley in South Dakota and northeastern Nebraska. The species is also known from the Pierre Shale in southwestern North Dakota and from the Salt Grass Member of the Pierre Shale of western Kansas.

Baculites baculus Meek and Hayden, 1861

Baculites baculus can attain a large size and has an elliptical to ovate cross section in young juveniles. As it matures, the cross section can become circular to almost square. It has nearly smooth flanks in juveniles, but as they increase in size, they develop low, broad lateral ribs or undulations. *Baculites baculus* has a very slight degree of taper except in the smaller specimens. The suture pattern is simplified like that of pre-*gregoryensis* baculites. The species appears to be a migrant from the Gulf Coast region and was the ancestor of later endemic species. Its ancestor seems to be *Baculites undatus* that occurs in rocks of a much earlier time within the Western Interior. *Baculites baculus* has been found from Colorado to eastern Montana (Glendive area) and also in Alberta and Saskatchewan.

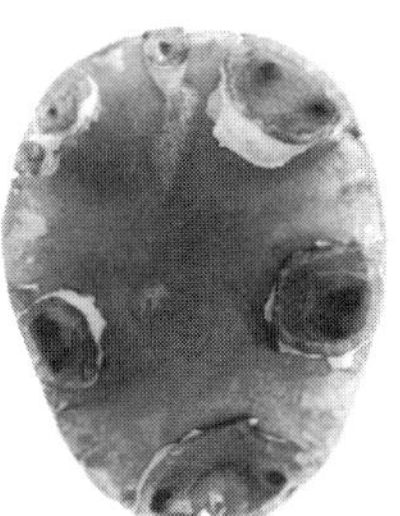

Left: Side view
Baculites baculus
BHI 4166
13.5 cm long
Above:
cross section

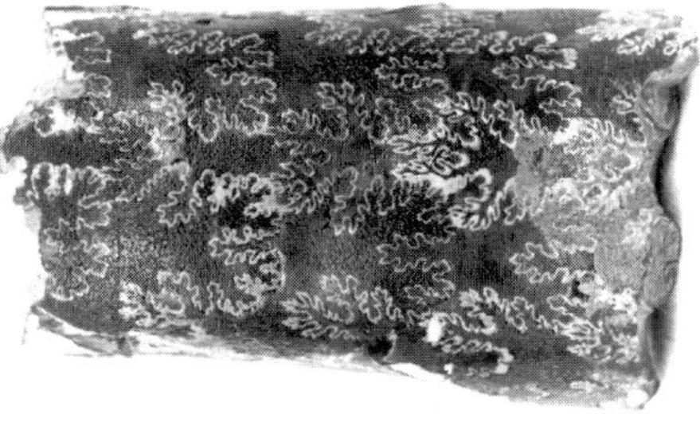

Baculites baculus
Above: BHI 4681
7.5 cm long
Below: BHI 4680
7.6 cm long

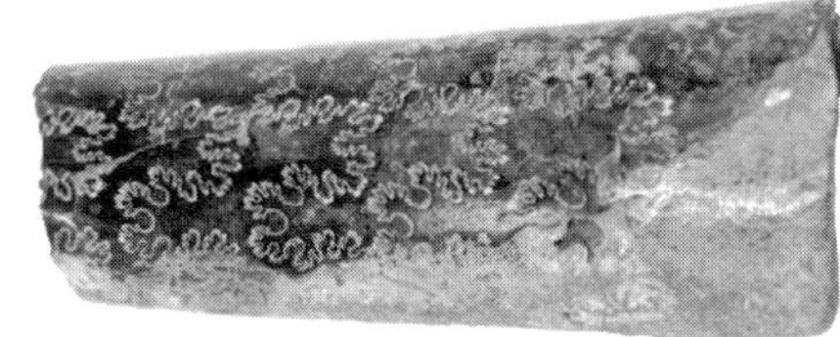

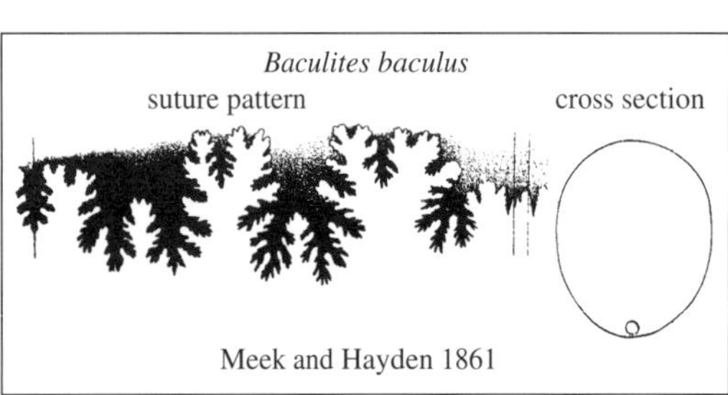

Baculites baculus
cross sections
Left: BHI 4680
Right: BHI 4681

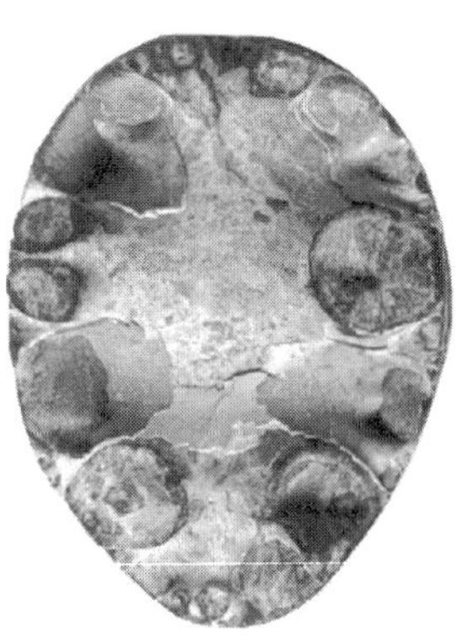

Baculites grandis Hall and Meek, 1854

This species name is derived from the extremely large size of mature adults. Some specimens of *Baculites grandis* have been confirmed at more than 100 cm (over 3 ft) long, making this species one of the largest known within the Family Baculitidae in the world. The cross section changes from nearly ovate in juveniles to almost trigonal in large adults. The flanks have very broad crescentic undulations, but smaller individuals do not show these undulations. The venter becomes more pronounced or narrow in relationship to the robust flanks and dorsum as the baculite matures. The smaller the specimen, the greater the taper. More complete specimens also show slight curvature. The suture pattern is similar to that of *Baculites baculus. Baculites grandis* ranges from Colorado and Kansas, and north into Montana.

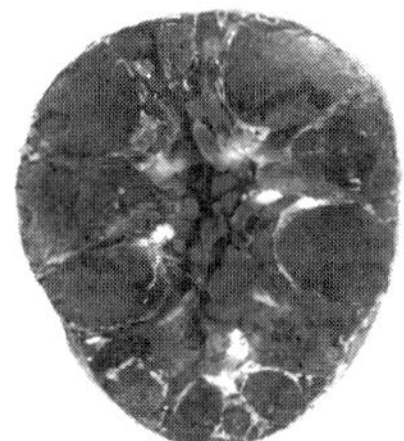

Baculites grandis
BHI 4171
cross section

Baculites grandis
BHI 4171
ventral view
16 cm long

Baculites grandis
S Jorgensen
collection
84 cm long

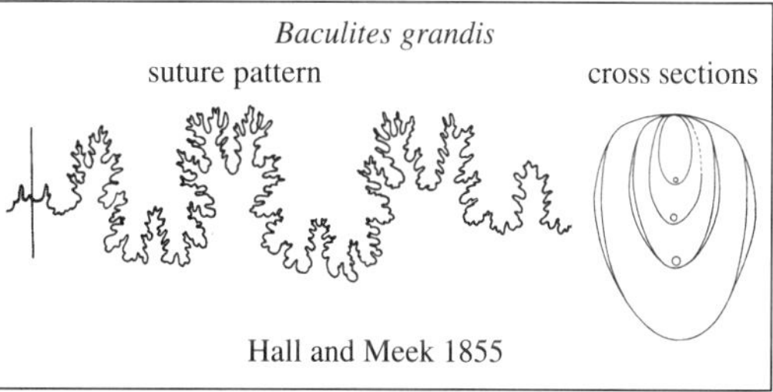

Baculites grandis
suture pattern   cross sections

Hall and Meek 1855

Baculites clinolobatus Elias, 1933

These are the last of the large baculites from the Western Interior. The cross section is compressed ovate in younger conchs to more triangular in large adults. The angle of taper is low. Larger adults developed widely spaced, low, broad undulations on the flank, whereas young individuals are smooth. *Baculites clinolobatus* has a more complex suture pattern than *Baculites grandis* and somewhat resembles the suture of *Baculites gregoryensis*. The species has been found in Kansas, Colorado, Wyoming, and South Dakota.

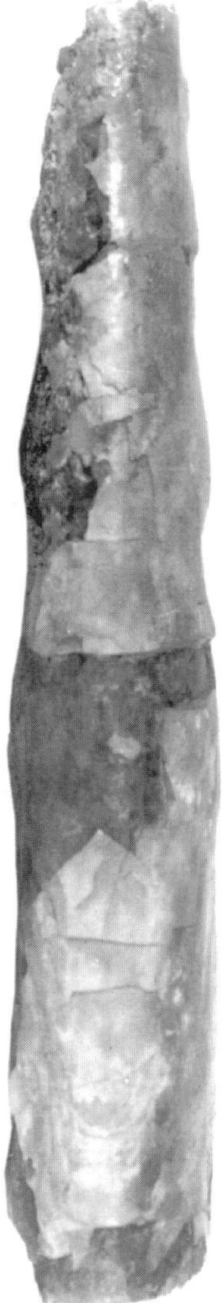

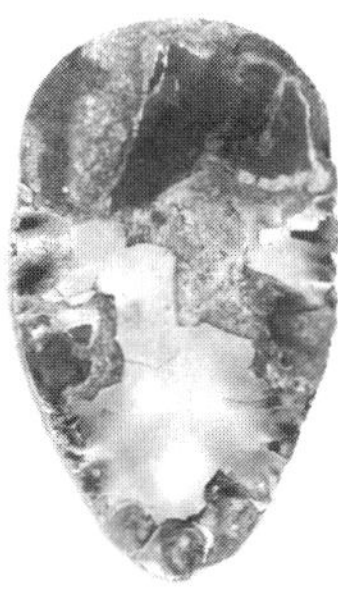

Left:
Baculites clinolobatus
DMNH specimen
(adult)
28 cm long

Below:
cross section

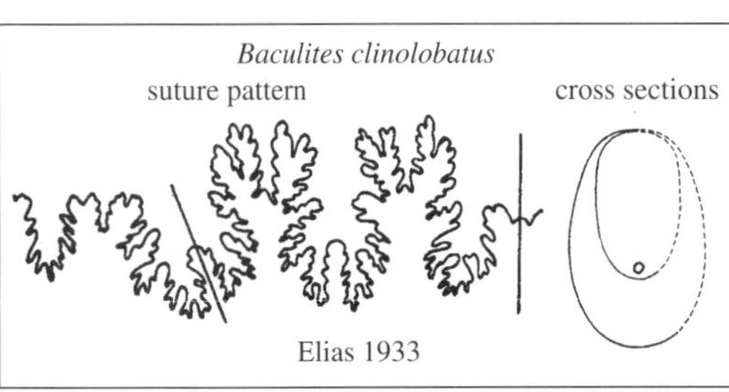

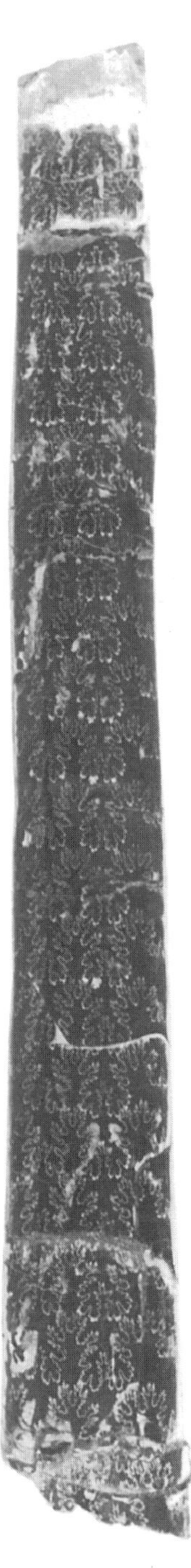

Baculites clinolobatus
Above: BHI 4759
12.5 cm long
Right: BHI 4758
20 cm long

Genus *Pseudobaculites* Cobban, 1952b

pseudo = false + *baculus* = a staff + *ites* = a stone

This genus has a straight shell with a rapid taper of generally 10° to 13°, an oval to elliptical whorl section, and broad undulations on the venter. Individual specimens are shaped more like flattened cones than representatives of the Family Baculitidae. The flanks are smooth to very ornate, with some forms having nodes on the dorsal slope of the flank. The genus differs dramatically from its contemporaries by having an unusually large shell with a very rapid taper and a complex suture pattern similar to *Baculites compressus*. *Pseudobaculites* has been found in rocks of Late Coniacian Age in Wyoming, Utah, and Colorado, and late Campanian rocks of Wyoming and Saskatchewan.

Pseudobaculitites natosini
USGS D4508
24 cm long

Pseudobaculites natosini (Robinson, 1945)

This species has perhaps the largest body chamber of any other species within the Family Baculitidae, yet with its rapid taper it never achieves a long size.

Pseudobaculites natosini has been described from the *Baculites jenseni* Zone of the Lewis Shale in central Wyoming, the *Baculites eliasi* Zone of the Pierre Shale in east central Wyoming, the Bearpaw Shale in Montana, the Pierre Shale of north central Colorado, and the Bearpaw Shale of southwestern Saskatchewan (Cobban and Kennedy, 1994b).

Pseudobaculitites natosini
USGS D4508 cross section

The authors have compiled this guide in the hopes of simplifying the process of identifying baculite specimens. The authors intend that this identification guide should be used in conjunction with the preceding scientific descriptions. Please refer to them for additional information.

The baculites are described as small, moderate, or large. A moderate-sized baculite is one that has a cross section of 2 to 5 cm or 1 in. by 2 in. However, smaller and larger specimens of the same species are found because of the presence of juveniles and large adults.

We have also described ventral and flank ornamentation with the brief descriptions of the baculites. These forms of ornamentation are rarely present on juvenile specimens. Nearly all mature forms of baculites develop some sort of flank ribs, and may also lose their ventral ribs. Very old baculites may lose all ornamentation and become more inflated.

The cross sections are also quite variable. Young, mature, and old baculites of the same species will vary to some degree. Different localities within the same Range Zone might also produce slightly different ornamentation and cross sections within the same species.

A low degree of taper is considered by the authors to be less than 2°, and a high degree of taper is 4° or more. Please consult the references if you have other questions concerning identification.

The following identifications are organized from youngest to oldest form.

Typical suture pattern

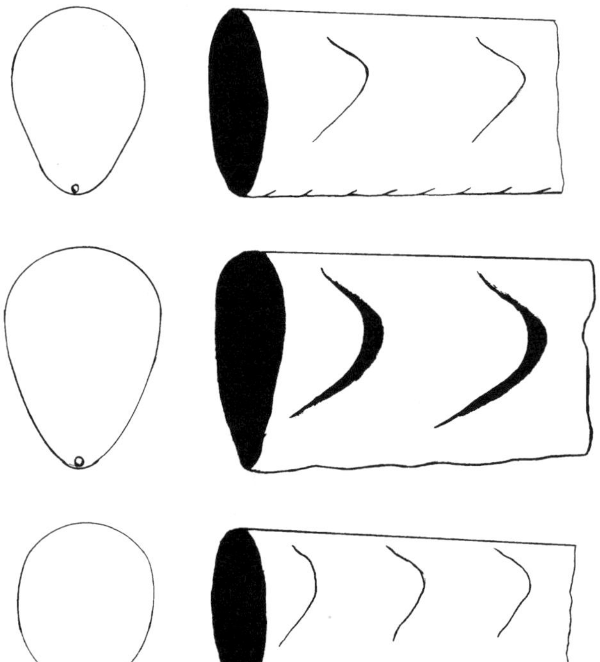

Baculites clinolobatus Elias, 1933
Elliptical to compressed trigonal cross section
Generally smooth venter
Smooth flanks developing undulations
Low degree of taper
Moderate to large

Baculites grandis Hall and Meek, 1854
Ovate to nearly trigonal cross section
Smooth to broad ventral ribs
Generally broad undulations with a
ventrolateral depression
Low to moderate degree of taper
Large

Baculites baculus Meek and Hayden, 1861
Ovate to circular in cross section
Smooth to well-ribbed venter
Smooth to undulated flanks
Low degree of taper
Large

Illustrations by N. Larson

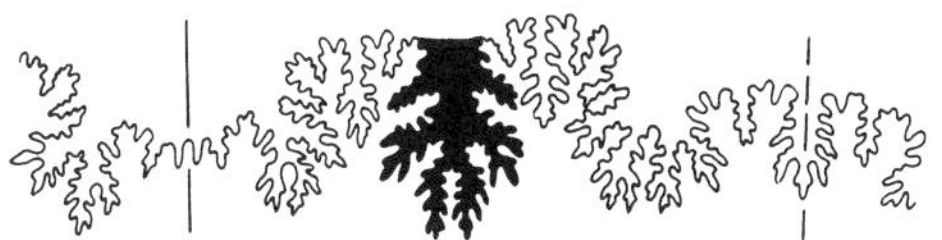

Typical suture pattern

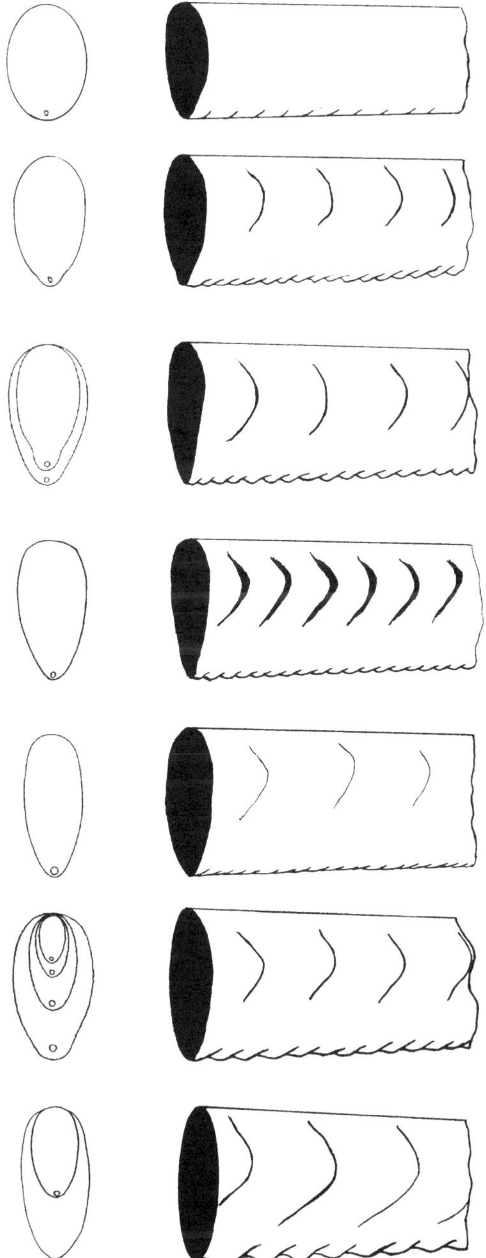

Baculites eliasi Cobban, 1958a
Ovate cross section
Smooth venter
Smooth flanks
Low degree of taper
Moderate

Baculites jenseni Cobban, 1962a
Strongly ovate cross section
Ventral ribs averaging nine to twelve per shell diameter
Smooth flanks with ventrolateral depression
Low degree of taper
Moderate

Baculites reesidei Elias, 1933
Compressed cross section
Ventral ribs averaging seven to eight per shell diameter
Smooth to weakly ribbed flanks
Moderate degree of taper
Moderate to large

Baculites cuneatus Cobban, 1962a
Trigonal cross section
Strong ventral ribs, six to nine per shell diameter
Broad short lateral ribs, one to three per shell diameter
High to low degree of taper
Moderate to large

Baculites compressus Say, 1820
Compressed cross section
Smooth venter
Smooth to weakly ribbed flanks
High to low degree of taper
Moderate to large

Baculites corrugatus Elias, 1933
Compressed cross section
Prominent ventral ribs, five to seven per shell diameter
Smooth flanks
Slight degree of taper
Moderate to large

Baculites rugosus Cobban, 1962a
Compressed cross section
Strong ventral ribs, three to four per shell diameter
Smooth to weakly ribbed flanks
High to low degree of taper
Moderate to large

Typical suture pattern

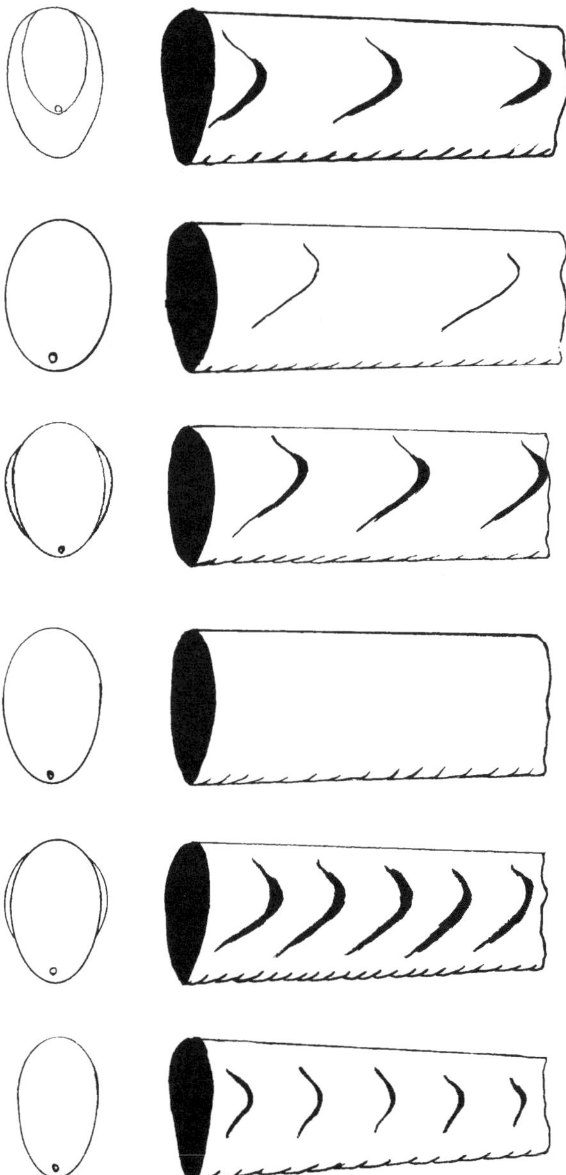

Baculites crickmayi Williams, 1930
Ovate cross section
Smooth to well-ribbed venter
Sculpted flanks with broad undulations,
about one per shell diameter
Gentle taper
Moderate

Baculites pseudovatus Elias, 1933
Ovate cross section
Smooth to weakly ribbed venter
Smooth to weakly ribbed flanks
Gentle taper
Moderate

Baculites sp. (new species)
Ovate cross section
Nearly smooth venter
Broad, almost nodelike ribs,
one per shell diameter
Gentle taper
Moderate

Baculites scotti Cobban, 1958a
Compressed ovate cross section
Smooth venter to slightly ribbed
Smooth to slightly ribbed flanks
Gentle taper
Moderate

Baculites reduncus Cobban, 1977
Ventrally compressed ovate cross section
Smooth to slightly ribbed venter
Broad arcuate flank ribs,
about two per shell diameter
High degree of taper
Moderate

Baculites gregoryensis Cobban, 1951a
Compressed elliptical cross section
Slight to well-ribbed venter
Smooth flanks on juveniles
Slightly ribbed flanks on adults
High degree of taper
Moderate

Typical suture pattern

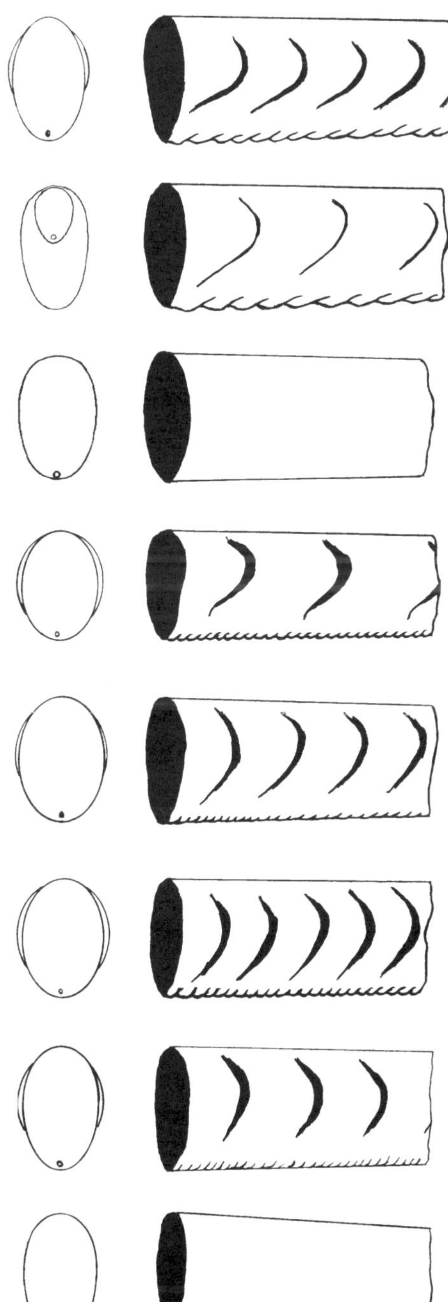

Baculites gilberti Cobban, 1962b
Subelliptical cross section
Prominent ventral ribs, about six per shell diameter
Broad flank ribs, about two per shell diameter
Low degree of taper
Moderate

Baculites perplexus Cobban, 1962b
Elliptical cross section
Prominent ventral ribs, about four per shell diameter
Smooth to weakly ribbed flanks
Low degree of taper
Moderate

Baculites sp. (smooth)
Subelliptical cross section
Smooth venter
Smooth flanks
Low degree of taper
Moderate

Baculites asperiformis Meek, 1876
Ovate cross section
Smooth to well-ribbed venter
Strong, oblique undulations or broad ribs, one per shell diameter Low
degree of taper
Moderate

Baculites mclearni Landes, 1940
Moderate to ovate cross section
Smooth to weakly ribbed venter
Widely spaced, broad flank ribs, about 1 1/2 per shell diameter Low
degree of taper
Moderate

Baculites obtusus Meek, 1876
Ovate cross section
Smooth to well-ribbed venter
Nodelike flank ribs, about two per shell diameter
Low degree of taper
Moderate

Baculites sp. (weak flank ribs)
Round to ovate cross section
Minor to well-ribbed venter
Broad arcuate ribs or swellings on flanks,
about one per shell diameter
Moderate

Baculites sp. (smooth species)
Ovate cross section
Minor ribbing on venter
Smooth flanks
Slight degree of taper
Moderate

The Family Collignoniceratidae is characterized by compressed oval to nearly square whorled forms of planispiral ammonites with a generally evolute umbilicus. The venter has a serrated or notched keel with clavate-type nodes and tubercles or clavi along the ventrolateral shoulder. Ribs are dense to widely spaced, generally broad, and typically ornamented with tubercles, clavi, or bullae. The family is distributed worldwide from the Lower Turonian through the Middle Campanian.

Genus *Menabites* (*Delawarella*) Collignon, 1948

Menabe = a town in Madagascar + *ites* = a stone

This genus and subgenus have moderately ovate to subrectangular whorl sections. Ornamentation consists of prominent umbilical and ventral shoulder tubercles on every rib. On adult forms, the ribs become more faint and generally display one or two rows of tubercles. The ventral edge is slightly depressed on either side of the keel when it again forms a rounded ridge above the siphuncle. The genus and subgenus have been reported from France, Tunisia, Madagascar, New Jersey, and occasionally within the Western Interior of North America.

Menabites (*Delawarella*) *danei* (Young, 1963)*

Menabites (*Delawarella*) *danei* is distinguished by trituberculate ornamentation, low, broad ribbing, an evolute umbilicus, and an oval to subrectangular whorl section. The venter has a broad, shallow groove separating the ventral shoulder from the siphuncle ridge. There are nine to ten ribs per whorl section that are prorsiradiate, broad, and coarse, with wide spacing. Tubercles are present on the umbilical shoulder, midflank, and on the ventral edge. Clavi are present on either side of the siphuncle ridge. The species has been found in the Gober Chalk and the Ozan Formation of Texas, the Ozan Formation of Arkansas, and is sparsely reported from the *Baculites obtusus* Zone of the Apache Creek Sandstone Member of the Pierre Shale near Pueblo, Colorado.

Side and ventral view
Menabites danei
USNM 14571
13.3 cm high

* Parentheses enclosing the author's name indicate that his original description placed the species in a genus not currently accepted for that species.

Menabites (Delawarella) vanuxemi (Morton, 1830)

This species can be quite large (up to 17.5 cm), and it is generally compressed with a sub-rectangular whorl section. The umbilical wall is short and steep, and the flanks are flat to slightly rounded. Ribbing is dense with 22 to 23 umbilical bullae per whorl, giving rise to 35 or 36 straight, prorsiradiate ribs. All the ribs bear umbilical and ventral shoulder tubercles, midflank tubercles, and ventral, marginal clavi. As in other *Menabites,* a small shallow groove separates the ventral marginal clavi from the siphuncle ridge. This species is present in New Jersey, Texas, and the upper part of the Mancos Shale in New Mexico.

Side and venter view
Menabites vanuxemi
USNM 14543
12 cm high

Genus *Submortoniceras* Spath, 1926
sub = lower in rank + *S. G. Morton* + *ceras* = horn

The description for this genus closely follows that of the family. Characteristics specific to this genus are a compressed whorl section, broad, rounded, inner flanks, flat outer flanks, and a flat venter bordered by tubercles with a slight, rounded ridge over the siphuncle. The flanks have low, broad ribs with lateral, ventrolateral, and umbilical shoulder tubercles. The species, *Submortoniceras tequesquitense* (Young, 1963), has been found in the upper Mancos Shale of New Mexico. The genus has also been found in Africa, Japan, Delaware, New Jersey, and Texas. We unfortunately do not have a photograph of this species.

The Family Desmoceratidae is defined by thick, round flanks and planispiral coiling. The whorls are generally round, oval, or even spear-shaped (having pronounced keels). The flanks are generally smooth to weakly ribbed with occasional constrictions. The family occurs worldwide from the Lower Cretaceous through the Upper Cretaceous.

Genus *Parapuzosia* Nowak, 1913

para = near + *puzosia*

The genus commonly attains a size greater than 60 cm. The umbilicus is moderately involute, and the umbilical walls are steep and high. Flanks are flat to slightly rounded with prominent primary and some secondary ribs on the inner whorls. The genus occurs from the Cenomanian through the Campanian in Africa, Europe, and North America.

Parapuzosia bradyi Miller and Youngquist, 1946

This species is the largest known of all the ammonite species from North America. With verified reports of more than 137 cm (4.5 ft) across and estimations of greater than 180 cm (6 ft) across, this ammonite was truly a giant. The whorl section is oval shaped; the umbilicus is moderately involute with steep, high walls. The flanks are generally smooth and rounded with ribs present on the inner whorls. The species has been found only in the upper part of the Eagle Sandstone of Big Horn County, Montana, and the upper part of the Cody Shale (*Scaphites hippocrepis* Range Zones), Big Horn County, Wyoming.

Parapuzosia bradyi

Carol Cheatham, a rockhound from Greybull, Wyoming, is shown standing next to the largest known ammonite from North America, *Parapuzosia bradyi*. The photo is dated 1964. This specimen was collected from the Cody Shale near Greybull and is on display at the Greybull Museum. The specimen is missing at least one-half whorl of body chamber. If complete, its size would increase to over 6 feet across. Note also the "smaller" specimen on the right. Photo courtesy of Dr. William Cobban.

Diplomoceratidae seems to be an offshoot of the Nostoceratidae family. They have loose coils tending to bilateral symmetry in one plane and develop fine to broad ribs with constrictions at different stages of growth. Some forms had ventrolateral tubercles or spines. However, most varieties did not possess this form of ornamentation. The family occurs worldwide in the Upper Cretaceous from the Turonian through the Maastrichtian.

Genus *Exiteloceras* Hyatt, 1894

exitel = becoming extinct + *ceras* = horn

This genus is described as having loose, elliptical, planispiral coiling and a subcompressed to oval-shaped whorl section. Ribs are simple in the early stages, becoming more complex with branching and intercalation in later stages. Nearly all of the ribs that cross the flanks and venter have tubercles and spines. The venter can be flat to rounded and the suture pattern is complex. The genus occurs in the Campanian of the Western Interior Seaway and the Mount Laurel Sand of Delaware. There is also a report of it possibly having been discovered in Colombia, South America.

Exiteloceras jenneyi Whitfield, 1877

Exiteloceras jenneyi is the only described species for the genus, so the description follows that of the genus very closely. The species in the juvenile stage resembles a juvenile baculite with the small ammonitella followed by a nearly straight, slightly ribbed shaft and angular to semicircular bends and loosely coiled limbs not in contact with the adjacent whorls. Ribs are moderately coarse, generally rursiradiate, and most contain ventrolateral spines and tubercles. *Exiteloceras jenneyi* has been found in the Western Interior in the Pierre Shale of Montana, South Dakota, Wyoming, Colorado, and New Mexico; the Bearpaw Shale of Montana; the Mancos Shale and the Iles Formation of Colorado; and the Lewis and Pierre Shales of New Mexico.

<table>
<tr><td align="center">Exiteloceras jenneyi
BHI 4149
7.6 cm across</td><td align="center">Exiteloceras jenneyi
BHI 4115
17 cm wide, 14 cm high</td></tr>
</table>

Genus *Glyptoxoceras* Spath, 1925

glypto = curved + *ceras* = horn

This genus begins with an initial shallow coil (similar to baculites) followed by either loose elliptical to round coils within one plane or a helical spire of several whorls. The shell is small and apparently does not exceed 10 cm in length. The whorl section is oval, and the ribbing is generally rectiradiate and sharp. Suture patterns are quite simple and the absence of nodes or tubercles make this genus quite distinctive.

Glyptoxoceras rubeyi (Reeside, 1927)

The specimens of *Glyptoxoceras rubeyi* (Reeside, 1927) from the Pierre Seaway seem to have an absence of ribs in the early whorls, more distinct ribs in the middle whorls, and some specimens have almost an absence of ribs on the last portion of the shell. They have been reported from the lower half of the Pierre Shale on the western rim of the Black Hills, from the Steele Shale of central Wyoming, and from the upper part of the Mancos Shale in Moffat County, Colorado.

Glyptoxoceras rubeyi
USNM 73292
2.3 cm across

Regarding this species, W. A. Cobban (written communication, 1994) states: "This genus is in need of revision inasmuch as it includes species that have similar adult growth stages and ornament but very different juvenile growth stages. Specimens from the Gammon Ferruginous Member of the Pierre Shale (*Scaphites hippocrepis III*) begin with a minute planispiral coil (similar to that of *Baculites*) followed by loose elliptical to circular whorls. Specimens from north Texas, about the age of the *Baculites obtusus* Zone, begin as a small helix followed by loose circular whorls. Specimens from France (possibly about the age of *Baculites mclearni* or *Baculites asperiformis* Zone) begin as a minute coil (like that of *Baculites*) followed by a straight, smooth shaft, then a narrow helix, and finally the loose, circular to elliptical, adult coils."

Further research may find that *Glyptoxoceras* may represent more than one genus and that some may not belong in the Family Diplomoceratidae but perhaps in the Family Nostoceratidae.

Glyptoxoceras rubeyi
USGS D3647
5 cm across

Genus *Solenoceras* Conrad, 1860

solen = pipe + *ceras* = horn

Solenoceras begins with an ammonitella similar to baculites and continues with a long, straight shaft that at one point turns 180° and bends back on itself, similar to a bobbypin. The whorl section is ovate to circular. Ribs are quite distinct and generally lie perpendicular to or at slight angles to the shaft. Small tubercles and even spines occur on either side of the venter on nearly every rib. *Solenoceras* is found from Africa to Europe to North America in the Campanian and Maastrichtian Stages. Four species of *Solenoceras* have been described from the Pierre Shale: *Solenoceras crassum* (described following); *Solenoceras texanum* (Shumard, 1861) is known to occur infrequently from the *Baculites compressus* through the *Baculites reesidei* Zones of Colorado; *Solenoceras mortoni* (Meek, 1876) was described from the Big Bend of the Missouri, probably within the *Baculites gregoryensis* Range Zone; and *Solenoceras reesidei* (Stephenson, 1941) has been recorded from the *Baculites compressus* and *Baculites reesidei* Zones of northern Colorado. The *Solenoceras* found associated with *Didymoceras nebrascense*, *Didymoceras stevensoni*, and *Exiteloceras jenneyi* are, at the time of this writing, not published. The species found with *Didymoceras nebrascense* is pictured here as *Solenoceras* sp.

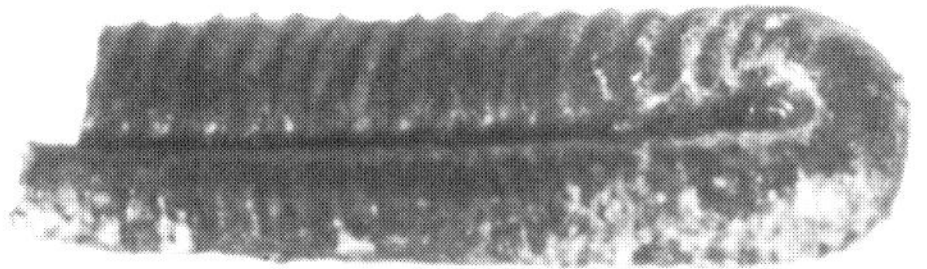

Solenoceras mortoni BHI 4019
1.5 cm long

Solenoceras sp. BHI specimen
7 cm long

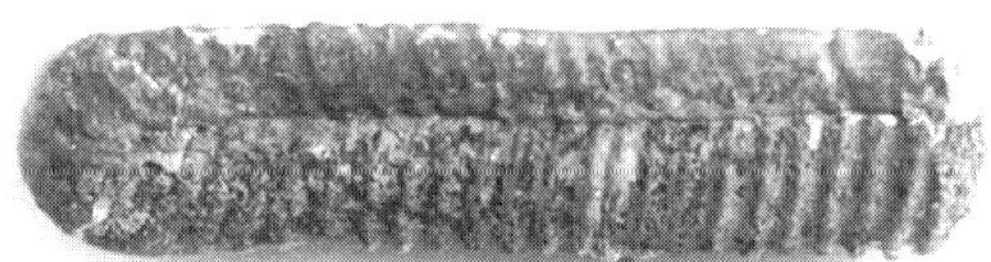

Solenoceras texanum
USGS D2825
3.1 cm long

Left:
Solenoceras sp. with *Baculites pseudovatus*
BHI 2146
Baculite is 6.5 cm long

Solenoceras crassum (Whitfield, 1877)

Solenoceras crassum follows the description of the genus. It is, however, quite large for the species, having been recorded up to 18.5 cm in length. It begins with an ammonitella, an open planispiral coil, a broadly curved limb, two parallel, tightly joined adult shafts joined by a tightly coiled, acute elbow, and rursiradiate ribbing on the body chamber. Each rib bears a tubercle on either side of the venter. Occurrences in the Pierre Seaway are from the *Didymoceras stevensoni* through the *Exiteloceras jenneyi* Range Zones from South Dakota, Wyoming, Colorado, and Montana. It has also been recorded from Colombia, South America.

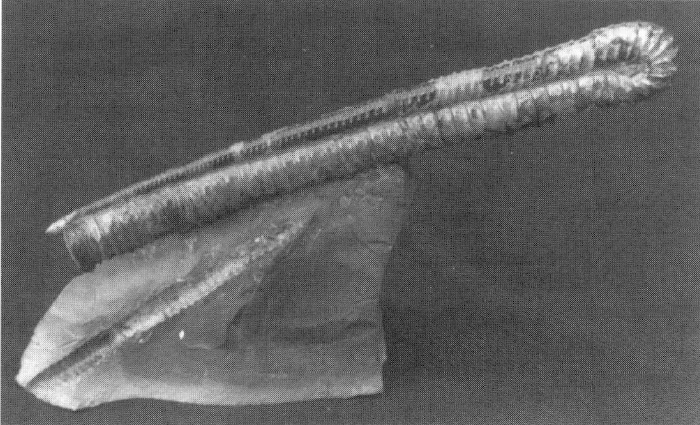

Solenoceras crassum
Didymoceras stevensoni Zone
17 cm long
Japh Boyce photo

Genus *Parasolenoceras* Collignon, 1969
para = alongside + *solen* = pipe + *ceras* = horn

Parasolenoceras is characterized by two widely separated, small parallel shafts connected by an elbow. The initial shaft, consisting of the phragmocone, is rectangular to circular in whorl section with low distinct ribs. The elbow and second shaft usually encompass the body chamber and are circular to compressed in whorl section. Ribs are generally prorsiradiate on the phragmocone, crowded to rursiradiate at the elbow, and straight to prorsiradiate on the body chamber. Some tubercles are present on either side of the venter, although they are small. One species, *Parasolenoceras pulcher,* Cobban and Kennedy, 1991a, was found in the *Baculites reesidei* Zone in northern Colorado. The genus has also been reported from the Gulf Coast of the United States and Madagascar.

Parasolenoceras pulcher
USGS 476127
5.2 cm long

The Family Nostocratidae is described as helicoid forms of regular or irregular coiling that can occur at early or late growth stages or anytime throughout its growth. Tubercles and spines are common along the venter and can be paired, irregular, a single row, or completely lacking. Ribs are generally broad, narrow, straight, and dense. They may also be fine and curved. Nostoceratids may show signs of uniform constrictions during any point of their growth; this may be a result of growth stopping or slowing down for a short period of time. There are sinistral (left) coiling and dextral (right) coiling forms for nearly all the known species of the family. They apparently descended from the Albian Stage *Turrilitoides* from England. The family occurs from the Cenomanian through the Maastrichtian Stage.

Life reconstruction of a
Didymoceras nebrascense
Family Nostoceratidae
Illustration by D.S. Norton

Genus *Anaklinoceras* Stephenson, 1941

anaklin = lean on + *ceras* = horn

This small genus has an initial tapering spire of whorls that lies within the open body chamber. The ribbing is strong and prominent, highlighted by small tubercles, mostly on the body chamber. Because of its small size, this genus may often be overlooked when concretions are being inspected. The genus has been found in Israel, Texas, Colorado, New Jersey, and possibly also in South Dakota. Two species have been named, and an as yet unpublished species has been reported from the *Baculites scotti* Range Zone of Colorado.

Anaklinoceras gordiale Kennedy and Cobban, 1992

Anaklinoceras gordiale has been found only in the *Baculites compressus* Zone of the Pierre Shale in Grand County, Colorado. It consists of a small spire of about $3^{1}/_{2}$ coils surrounded by the body chamber that is in a planispiral coil. The ribs are widely spaced, rounded, and rectiradiate in the early whorls and rursiradiate for the remaining ribs. This species is small and only reaches 2 cm in size.

Anaklinoceras gordiale
USNM 449783
2 cm across

Anaklinoceras reflexum Stephenson, 1941

This species achieves a slightly larger size than *Anaklinoceras gordiale*, up to 3.5 cm for macroconchs and only about 2 cm for microconchs. The phragmocone has about 4 to 4^1/$_2$ coils set in a tapered spire and an inverted J-shaped body chamber that wraps around the initial coils. Ribbing is inconspicuous in the early stage, but becomes stronger and more prominent throughout the growth of the shell. The venter is somewhat flat, and ventrolateral bullae and tubercles are present throughout most of the coiling. The species has been found in the *Baculites compressus* Zone of the Pierre Shale, Grand County, Colorado; Texas, New Jersey, and Israel.

Front and rear view
Anaklinoceras reflexum
USNM 445111
4.7 cm high

Genus *Axonoceras* Stephenson, 1941

axon = axle + *ceras* = horn

The genus *Axonoceras* is determined by small, loosely coiled, symmetrical whorls within one plane and small, simple ribs with ventrolateral tubercles. *Axonoceras* is known from the Campanian of Angola, and Texas, Mississippi, and Colorado.

Axonoceras compressum Stephenson, 1941

This species is a small (less than 3 cm) planispiral ammonite with approximately three slightly  separated whorls. The whorl section is elliptical to ovate, and the ornamentation consists of numerous narrow, prorsiradiate, flexuous ribs, many of which have tubercles or clavi as they cross the ventrolateral shoulder. The venter is flat, and the suture pattern is simple. The species has been described from the *Baculites compressus* Zone of the Pierre Shale in the Kremmling area of Colorado.

Axonoceras compressum
USGS D8093
2 cm across

Genus *Cirroceras* Conrad, 1868

cirros = curl + *ceras* = horn

Cirroceras is described as having several helical whorls, not touching, and with a loose U-shaped body chamber. It has dense, strong ribs and tubercles along the flattened venter in either the entire length of the shell or only on the body chamber. The genus has been found from the Campanian and Maastrichtian Stages in western Europe, in Nigeria and Angola, and in the Gulf and Atlantic coasts of North America, British Columbia, and only occasionally within the Pierre Seaway. *Cirroceras conradi* (Morton, 1841) has been found in the Western Interior in the *Baculites cuneatus* Range Zone near Kremmling, Colorado. The species is listed as *Didymoceras conradi* by Cobban and Kennedy (in a yet unpublished manuscript), but those authors now believe the species, with its identical appearance to the Gulf Coast specimens, is better assigned to the genus *Cirroceras*.

Cirroceras conradi
USNM 16213
6.7 cm across

Genus *Didymoceras* Hyatt, 1894

didymos = double + *ceras* = horn

The genus *Didymoceras* as described by Hyatt consists of loose helical spires that have circular whorl sections, two rows of paired or irregular tubercles along the venter, irregularly bifurcated ribs, and a large U-shaped body chamber. The genus coils sinistrally or dextrally. *Didymoceras* has been found in the Campanian and Maastrichtian Stages of North America, Europe, the Middle East, and Africa.

Didymoceras tortum (Meek and Hayden, 1858)

This species has circular whorl sections and ribs that are weak on the dorsum and prorsiradiate, strong, coarse, and simple (though they may branch) on the flanks and venter. Tubercles are present on the larger whorls and bullae on the early whorls. There is a tendency for an irregular positioning of tubercles and a presence of nontuberculate ventrolateral ribs. Early growth has open helical whorls in contact with each other. Later growth stages suggest an opening or separation in their whorls from the earlier whorls, and maintain the open helicoid coiling. *Didymoceras tortum* is found in the *Baculites gregoryensis* Range Zone of South Dakota and Wyoming and in the Annona Chalk of Arkansas.

Didymoceras tortum
BHI 4108
7.5 x 11.5 cm

Didymoceras cochleatum (Meek and Hayden, 1858)

Didymoceras cochleatum is characterized by a circular whorl section and a slowly broadening, loosely coiled helicoid conch. Ribs are low, simple, and rursiradiate, with two rows of tubercles, one midventer, and the other low on the outer whorl (base of the venter). Not all of the ribs have tubercles on the venter, but most exhibit this ornamentation. *Didymoceras cochleatum* occurs in the *Baculites gregoryensis* Range Zone in central South Dakota and in the Annona Chalk of Howard County, Arkansas.

Didymoceras cochleatum
BHI - 4019
8 x 13 cm

Didymoceras mortoni (Hall and Meek, 1854)

Didymoceras mortoni differs from its contemporary *Didymoceras tortum* by its irregular positioning of tubercles on the ribs on the ventrolateral flank and by wider, stronger, straighter, and more distinct ribs on the flanks and venter. Initial coils are similar to *Didymoceras tortum*, but the whorls taper less or do not expand as rapidly as on *Didymoceras tortum*. Whereas *Didymoceras cochleatum* is loosely coiled, *Didymoceras mortoni* has narrow, loose whorls, and yet is tightly, helically coiled in early stages. Tubercles are midventer and low on the base of the outer whorl, and the ribs are weakly rursiradiate. *Didymoceras mortoni* occurs in the zones of *Baculites gregoryensis*, *Baculites reduncus*, and *Baculites scotti* of South Dakota, Wyoming, Colorado, and New Mexico, and in the Annona Chalk of Howard County, Arkansas.

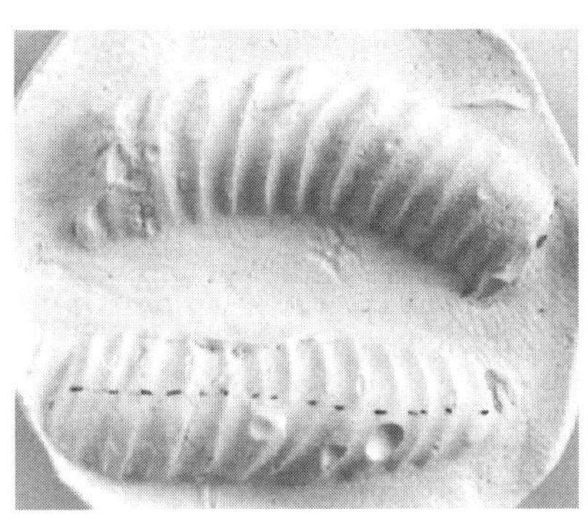

Didymoceras mortoni
cast of type
AMNH 9550
3 cm long

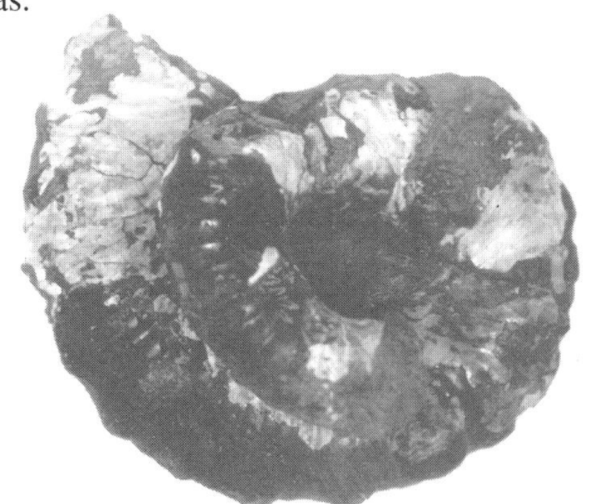

Didymoceras mortoni
BHI 4764
4 x 3.5 cm

Didymoceras binodosum (Kennedy and Cobban, 1993b)

This species begins with two open planispiral coils in contact, followed by a slightly curved slender shaft. The slender whorls then form a loose elliptical whorl that descends into two loose, circular, helical whorls on a low U-shaped body chamber. Ribs are prorsiradiate to rectiradiate on the early whorls and become rursiradiate on the lower two whorls and body chamber. This species exhibits obvious, unusually deep, periodic constrictions repeated every 25 to 30 ribs. There is a well-defined venter bordered by ventrolateral tubercles on most of the ribs. The later two coils tend to have tubercles or spines located slightly above midventer and another row of tubercles or spines low on the base of the venter. *Didymoceras binodosum* is found in the *Baculites scotti* Range Zone of the Western Interior in South Dakota, Wyoming, and Colorado; in the Annona Chalk near Yancy, Arkansas; and in the Bergstrom Formation in south central Texas.

Didymoceras binodosum
Di 487H
Steve Jorgensen collection
3cm across

Didymoceras binodosum
Di 588A
Steve Jorgensen collection
6 cm wide x 7cm high

Didymoceras cf. *archiacianum* (d'Orbigny, 1842)

This species is characterized by $1\frac{1}{2}$ to 2 loosely coiled helical whorls in its initial growth, followed by 5 to 6 helical whorls in contact with each other on a low, slightly separated body chamber. Ribs are generally strong and rursiradiate, weak to faint on the dorsum, and strong and broad on the flanks and venter. Ribs are sometimes paired across the flank, meeting to form tubercles or spines midway on the ventral edge, and looped between these tubercles and another row, low on the outer whorl. *Didymoceras* cf. *archiacianum* occurs in the *Baculites scotti* Range Zone in South Dakota, Wyoming, and Colorado, and shows distinct similarities to *Didymoceras archiacianum* from France. The authors believe enough differences do exist, however, and *Didymoceras cf. archiacianum* will one day be assigned a different species name.

Left specimen: *Didymoceras cf. archiacianum*
Di 984E 8.2 cm high

Right: *Didymoceras cf. archiacianum*
Di 1275B 13.5 cm high

S. Jorgensen collection

Didymoceras nebrascense (Meek and Hayden, 1856)

Didymoceras nebrascense typically consists of $3^1/2$ loosely coiled, broad, helical whorls on a large, broad, U-shaped body chamber. Early whorls consist of an ammonitella followed by a nearly straight limb that bends back and gently tapers into a slightly curved limb in a loose, elliptical, planispiral whorl. Middle growth is generally $2^1/2$ broad, almost circular, loose helical coils. The body chamber becomes quite robust as it forms a U. Ribs are prorsiradiate early, low, and faint, then become wider and more rursiradiate on the remaining whorls and body chamber. Ventrolateral tubercles or spines do not occur on every rib but seem to be on only every third or fourth one for most of the growth. The final one-half to two-thirds of the whorl before the body chamber lacks tubercles. Tubercles occur high and low on the ventrolateral shoulder early, but slowly drop to midventer and low on the whorls until they fade out on the lower whorl. The body chamber has prominent ventrolateral spines bordering the nearly flat venter. Macroconch body chamber height averages 15 cm, with an overall height of 20 to 25 cm. Microconch body chamber height averages 12 cm, with an overall height of 14.5 to 17 cm. Macroconchs tend to have larger and more robust phragmocones and body chambers than do the microconchs of the species. *Didymoceras nebrascense* is prevalent throughout the Western Interior from New Mexico through northeast Montana. It is also found in Delaware and in the Mishash Formation of Israel.

Above:
Didymoceras nebrascense with
Solenoceras sp., *Baculites
pseudovatus*, and *Scaphites gilli*
BHI 4634
40 cm long
25 cm high

Didymoceras nebrascense
BHI 4113
21.5 cm high

Didymoceras nebrascense
BHI 4112
7 cm across

Didymoceras stevensoni (Whitfield, 1877)

Didymoceras stevensoni is described, in early growth, as a loose, elliptical, or upside-down U-shape with an initially curved, gently enlarging shaft on two slightly smaller, loose helical whorls. These whorls grade into three broad helical whorls that are in contact with each other or nearly so on a broad, shallow U-shaped body chamber. Ribs are broad and rectiradiate to rursiradiate. On the earlier whorls and limbs, the rib–venter junction exhibits tubercles for almost every rib; later in the growth, nontuberculate ribs are common. The venter or whorl flank is nearly flat and has two rows of tubercles, one near the base of the whorl and the other at midflank of the ventral side. *Didymoceras stevensoni* differs from *Didymoceras nebrascense* by having a larger number of tighter coiled, helical whorls (a spire) in the conch, the presence of ventrolateral tubercles on all whorls, coarser and more widely spaced ribbing, and a body chamber shaft that does not extend very far below the whorls before the recurvature in the form of a large U-shaped hook. Macroconchs are one-half to two-thirds the size of the macroconchs. This species very closely resembles its predecessor, *Didymoceras* cf.

archiacianum. *Didymoceras stevensoni* is found in the Western Interior from Colorado, Wyoming, and Montana, as well as in Delaware.

Didymoceras stevensoni
Left: Microconch
Right: Macroconch
24.5 cm high
Japh Boyce photo

Didymoceras cheyennense (Meek and Hayden, 1856)

Didymoceras cheyennense is typically characterized as $2^1/_2$ to 3 slender, loose helical whorls on a large J-shaped body chamber that has a very long shaft preceding the bend. Macroconch height averages about 20 to 21 cm, and microconch height averages 14 to 15 cm. *Didymoceras cheyennense* begins with an initial ammonitella and $1^1/_2$ open, yet touching, planispiral coils on a loose, slightly curved, elliptical whorl. The middle helical whorls enlarge slowly and are almost circular and yet not touching. The body chamber is fairly broad, with an initial shaft that extends about $1^1/_2$ times the height of the initial whorls, then bends back to about one-half of that distance, and forms the aperture. The early ribs are singular, prorsiradiate, and/or rectiradiate, and possess a double row of tubercles on either side of a nearly flat venter. Ribs on the middle whorls are strong and oblique, generally paired, rursiradiate on the venter and prorsiradiate on the top of the whorl. The ventrolateral tubercles tend to be on every other rib on the nearly flat venter. The ribs are broad on the body chamber. They are prorsiradiate at first, then rectiradiate on the curve, and rursiradiate on the final part of the hook. The tubercles are generally paired but occasionally alternate on either side of the venter on the body chamber. This species is found within the Western Interior Seaway from northern New Mexico to central and eastern Montana. It is also present in the Mount Laurel Sand of Delaware.

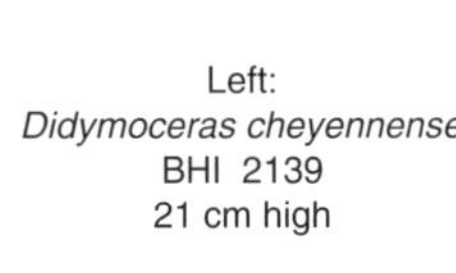

Left:
Didymoceras cheyennense
BHI 2139
21 cm high

Right:
Didymoceras cheyennense
BHI 2100
15 cm high

Genus *Nostoceras* Hyatt, 1894

nostos = return + *ceras* = horn

Nostoceras has acute angled, helical whorls in a tightly coiled spire on a loose U-shaped body chamber. Generally present are paired rows of tubercles on all or most ribs along the ventrolateral

Nostoceras monotuberculatum
BHI 2093
4.3 cm across

shoulder. The ribs are generally strong and dense with or without some constrictions. This Campanian and Maastrichtian genus has been found in Europe, Africa, North America, and Japan. Several species have been found within the Western Interior, but for the most part, they are very uncommon. A few specimens of *Nostoceras monotuberculatum*, Kennedy and Cobban (1993b) (so named because they have only one row of ventrolateral tubercles), have been found in the *Didymoceras nebrascense* Range Zone of Fall River County, South Dakota, and another specimen was found in the *Didymoceras stevensoni* Range Zone in the Cozzette Sandstone Member of the Mancos Shale, Pitkin County, Colorado.

Genus *Oxybeloceras* Hyatt, 1900

oxybel = eye of a needle + *ceras* = horn

At the time of this printing, the genus *Oxybeloceras* is in doubt and will most likely be changed. The type specimen for *Oxybeloceras* was *Oxybeloceras crassum,* and that specimen has been determined to actually be a large *Solenoceras.*

The initial coil of *Oxybeloceras* is similar to that of *Didymoceras cheyennense.* Early stages of *Oxybeloceras* consist of an ammonitella and 1 $^1/_2$ open planispiral whorls in contact with each other, followed by a long, slightly curved shaft that eventually bends back tightly on itself, leaving a small open tear-shaped area near the elbow. The long body chamber includes the elbow and the entire second shaft that bends back against the initial shaft. The ribs are generally straight and perpendicular to the venter and the dorsum. Small tubercles or even spines occur on the ribs along the ventrolateral shoulders. *Oxybeloceras* has been reported from the Western Interior and from Texas, Arkansas, and Delaware, as well as from Colombia, South America. In the Western Interior they have been found in *Baculites gregoryensis, Baculites scotti,* and *Didymoceras cheyennense* Range Zones.

Oxybeloceras sp.
BHI 4111
Baculites scotti Range Zone
6.2 cm long

Oxybeloceras meekanum (Whitfield, 1877)

Oxybeloceras meekanum differs from the *Baculites scotti* Range Zone *Oxybeloceras* by having a much shorter body chamber ratio to the phragmacone size. The ribbing seems to be slightly less dense than on the earlier undescribed species. *Oxybeloceras meekanum* is also smaller in size than its predecessor. Following the initial coils, the slightly bent shaft broadens rapidly with rursiradiate ribbing throughout. Tubercles are present on all of the ribs on the ventrolateral shoulder. *Oxybeloceras meekanum* is found from northern New Mexico to Montana, including western Nebraska and South Dakota within the *Didymoceras cheyennense* Range Zone.

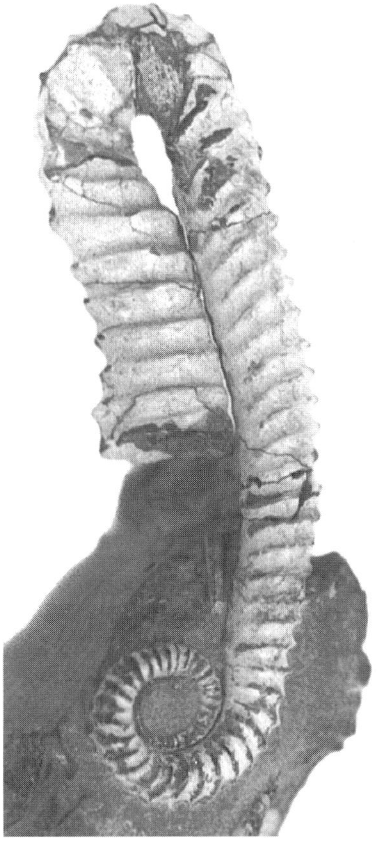

Oxybeloceras meekanum
BHI 4761
1.7 cm long
The initial coils of an *Oxybeloceras meekanum*. Photo is greatly
magnified, approximately 6.5 times lifesize.

Oxybeloceras meekanum
BHI 4114
5 cm long

Pachydiscidae includes small-, moderate-, and large-sized planispiral shells with a moderately, evolute to involute umbilicus. The whorl section can be inflated to compressed with generally steep umbilical walls. At early ontogenetic stages, they also tend to have strong rectiradiate ribs at some growth stage, and most forms generally have tubercles above the umbilicus along the umbilical shoulder. The family seems to have arisen in the Upper Albian and continued throughout the Maastrichtian Stage in all the shallow Cretaceous seas of the world.

Genus *Menuites* Spath, 1922
menu = small, detailed + *ites* = a stone

This ammonite shows the varied differences between the macroconch and the microconch. *Menuites* is characterized by moderately involute coiling, very broad, rounded flanks, a well-rounded venter, and deep, rounded umbilical walls. Strong, widely spaced ribs are present on the inner coils and at the aperture. Microconchs of the genus possess tubercles around the umbilical shoulder and ventrolateral tubercles on the living chamber, whereas the macroconch has smooth ventrolateral shoulders and either small umbilical tubercles or none. The genus has a worldwide distribution and has been noted in Africa, India, Japan, Europe, as well as the Western Interior in North America.

Menuites portlocki (Sharp, 1855) *complexus* (Hall and Meek, 1856)

Menuites portlocki complexus follows the description of the genus in having strong ribs (approximately 12 to 13 per one-half whorl in the phragmocone) that radiate from the umbilicus outward over the flanks and the venters, some connecting the tubercles and other ribs separating the tubercles. The macroconchs vary greatly with a larger size in mature adults and a lack of ornamentation with the exception of ribs on the phragmocone. Microconchs have prominent ribs and ventrolateral and umbilical tubercles. *Menuites portlocki complexus* is found mainly in the *Baculites gregoryensis* Zone in South Dakota, the Red Bird Silty Member of the Pierre Shale, the Rock River Formation in Wyoming, and the Hygiene Sandstone Member of the Pierre Shale in Colorado.

Menuites portlocki complexus
Top: Side view, Bottom: Venter view
Left: BHI 4144, 3.3 cm across
Right: BHI 4145, 3.3 cm across

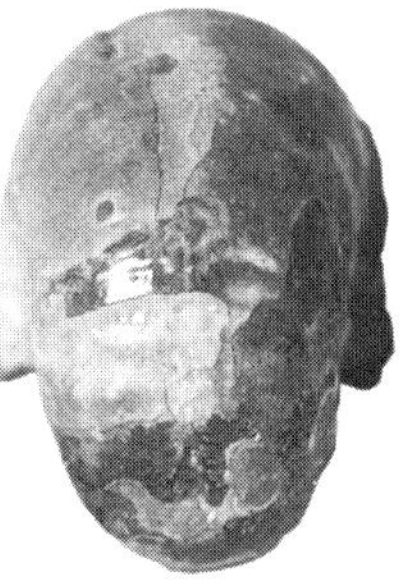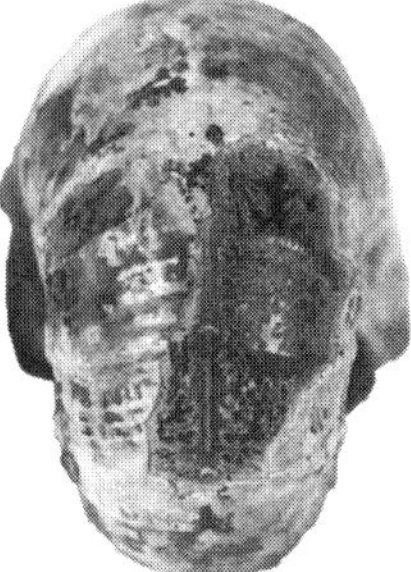

Menuites oralensis Cobban and Kennedy, 1993a

This ammonite also shows strong variations between the microconch and macroconch. The basic descriptions of this and *Menuites portlocki complexus* are the same, but *Menuites oralensis* has 15 to 30 ribs per one-half whorl. These ribs are difficult to see and irregular in height. On microconchs the ribs are weak or absent on the living chamber, except near the aperture. The living chamber also possesses ventrolateral tubercles and umbilical bullae along the umbilical shoulder that parallel or alternate with each other on both sides of the venter. The macroconchs get much larger (up to 30 cm) and lack ornamentation on the living chamber. *Menuites oralensis* occurs only in the *Baculites scotti* Range Zone of the Pierre Shale and equivalent-age rocks. The species has been reported from Colorado, South Dakota, Wyoming, and New Mexico.

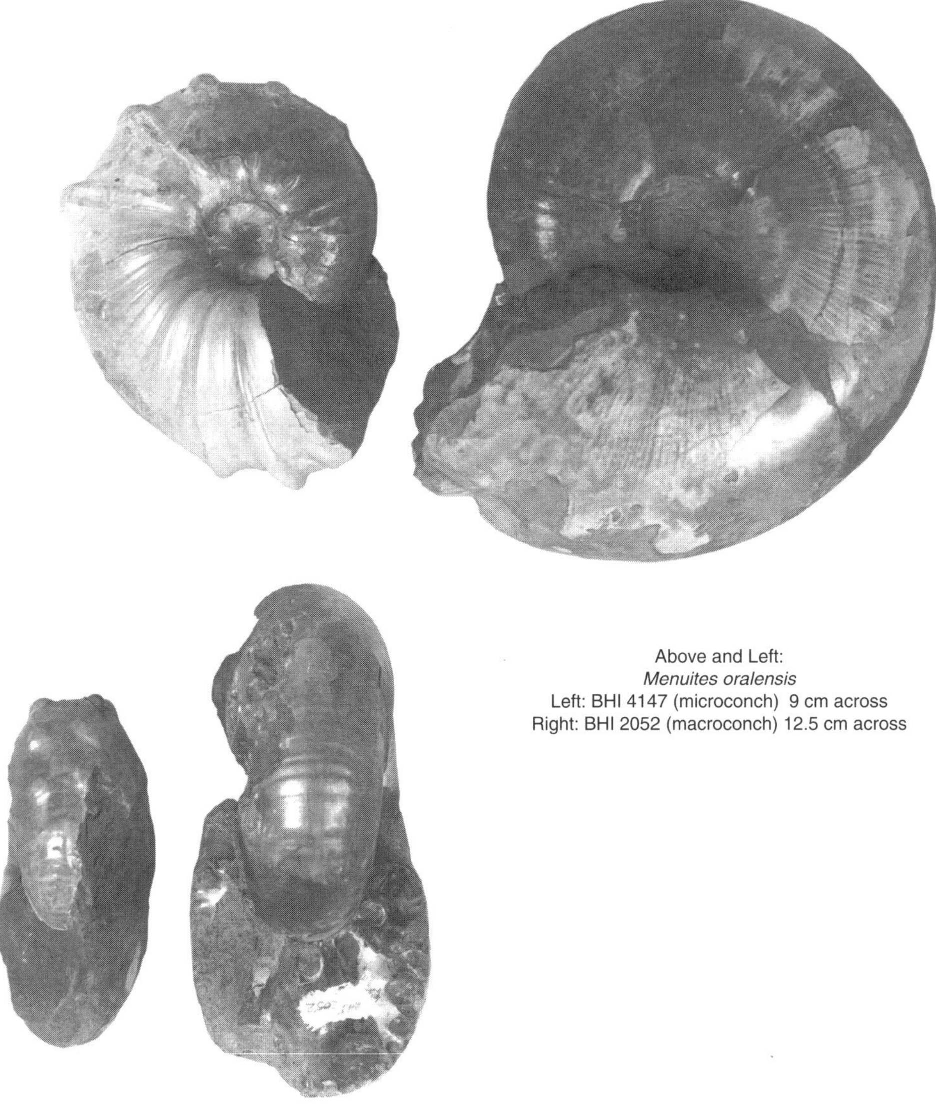

Above and Left:
Menuites oralensis
Left: BHI 4147 (microconch) 9 cm across
Right: BHI 2052 (macroconch) 12.5 cm across

Genus *Pachydiscus* Zittel, 1884

pachy = thick + *discus* = disk

The genus *Pachydiscus* is found worldwide. All the conchs within this genus have common characteristics of moderate coiling, slightly evolute, and well-rounded flanks blending into a rounded venter. The umbilicus is well rounded and very deep. There is usually strong ribbing at some stage of growth and a tendency for tubercles, especially around the umbilical margin. *Pachydiscus* has worldwide distribution from the Upper Albian through Maastrichtian Stages.

Pachydiscus (*Pachydiscus*) *arkansanus* (Stephenson, 1941)

This species is the most common form of large Pachydiscidae found within the Pierre Shale. Like all *Pachydiscus,* it is moderately evolute with rounded, deep umbilical walls. The flanks and venter are well rounded. *Pachydiscus* has strong, prominent ribs on the flank crossing the venter and strong primary ribs rising above the umbilicus on the umbilical shoulder. *Pachydiscus* (*Pachydiscus*) *arkansanus* can get quite large, probably greater than 60 cm across. It has been recorded from the *Exiteloceras jenneyi* Range Zone through the *Baculites compressus* Range Zone in the Pierre Shale. The significant difference between *Pachydiscus* (*Pachydiscus*) *arkansanus* and *Pachydiscus* (*Pachydiscus*) cf. *oldhami* is the more complex suture

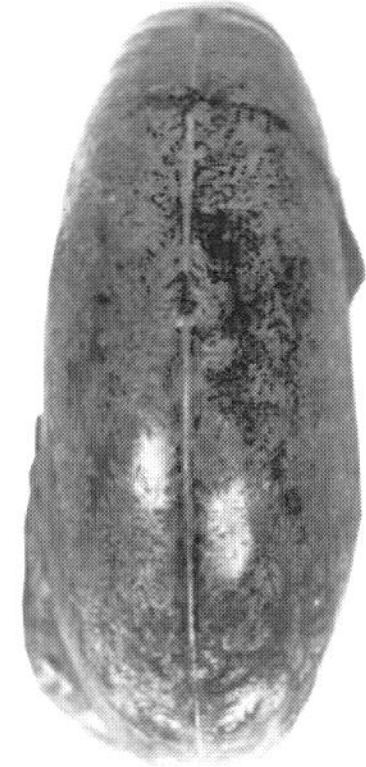

Side and edge view
Pachydiscus cf. hornbyense
BHI 4149, 23.5 cm across

pattern of *Pachydiscus* (*Pachydiscus*) *arkansanus*. *Pachydiscus* (*Pachydiscus*) cf. *oldhami* (Sharp, 1855) has been reported only from the Western Interior from the *Baculites compressus* Range Zone of the Bearpaw Shale in Montana. A single large phragmocone of *Pachydiscus* sp. was found by Walter G. Camack, Pueblo, Colorado, from the *Exiteloceras jenneyi* Range Zone in the Pierre Shale near Pueblo, Colorado (W. A. Cobban, written communication, 1994), and at least one specimen of *Pachydiscus catarinae* has been reported from the Pierre Shale near Pueblo, Colorado. A couple of specimens of *Pachydiscus* cf. *hornbyense* Jones, 1963 have been verified from the *Baculites compressus* and *Baculites cuneatus* Range Zones of Meade and Pennington Counties of South Dakota.

Cast of *Pachydiscus catarini*
USGS specimen
21 cm across

The Family Placenticeratidae is characterized by small to large, compressed, planispiral shells. The umbilicus is moderately involute, flanks are somewhat rounded, and the venter is flat or sometimes grooved. Most possess some form of ribbing, which tends to be slightly rectiradiate swellings, but many forms have no ribbing. Placenticeratidae have worldwide distribution from the Upper Albian through the Maastrichtian Stages.

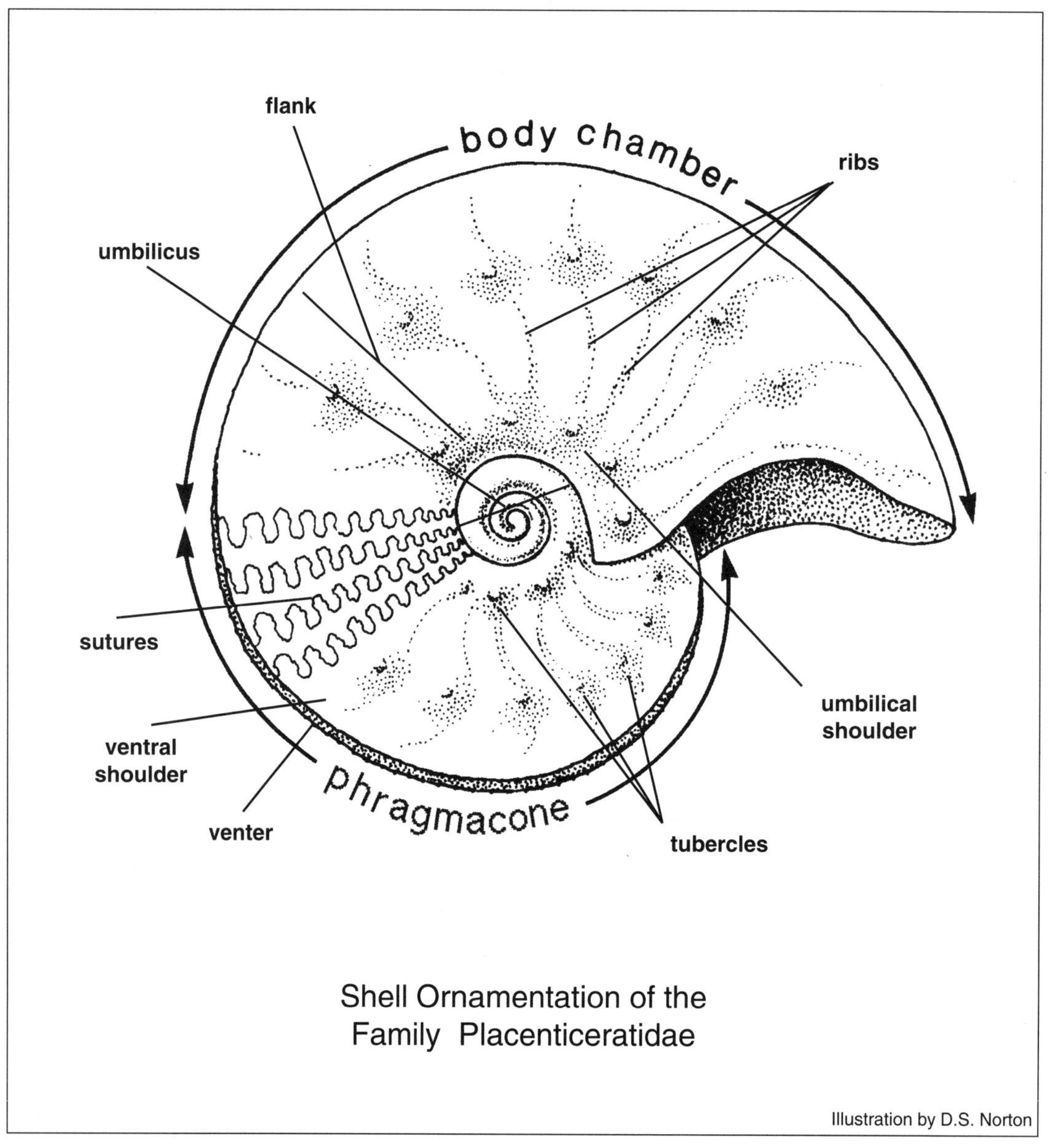

Shell Ornamentation of the
Family Placenticeratidae

Illustration by D.S. Norton

Genus *Hoplitoplacenticeras* Spath, 1922

hoplito = heavily armed foot soldier + *placenta* = flat cake + *ceras* = horn

Above: cast of
Hoplitoplacenticeras wesfieldensis
USNM 132373
8.2 cm high

This genus is characterized by a somewhat compressed whorl section, evolute umbilicus, flat to slightly rounded flanks, a flat venter, and prominent, straight to prosiradiate ribs that bear two rows of ventrolateral tubercles. The outer row of tubercles may be somewhat clavate. This genus is rare in the Western Interior. A single specimen assigned to *Hoplitoplacenticeras* cf. *Hoplitoplacenticeras coesfeldiense* (Schlüter, 1867) var. *schlüteri* Mikhailov, 1951, was found in the *Baculites asperiformis* Zone in the Steele Shale in south-central Wyoming. Two other specimens that seem to be *Hoplitoplacenticeras marroti* (Coquand, 1859) were found in the *Baculites* sp. (smooth) Zone and *Baculites obtusus* Zone in the Steele Shale and the Shannon Sandstone in central and southern Wyoming (W. A. Cobban, written communication, 1994).

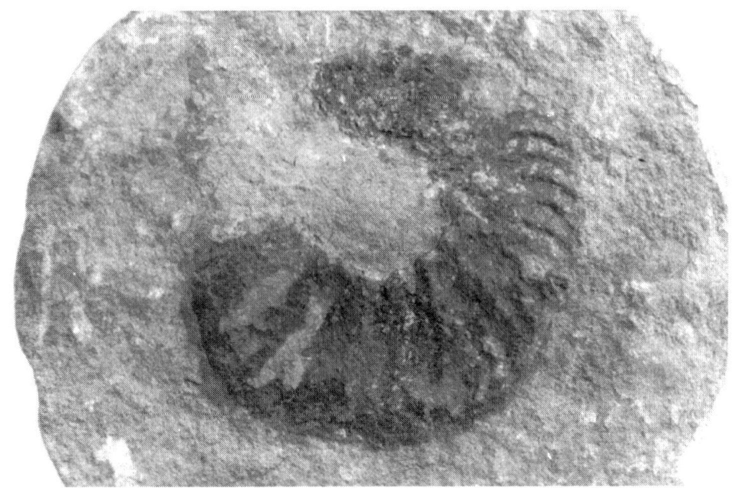

Above: cast of *Hoplitoplacenticeras marroti*
USGS specimen
Shannon Sandstone
10.8 cm high
This specimen was discovered in the center of a core
drill sample and is the finest representative of this
genus and species from the Pierre Seaway.

Genus *Placenticeras* Meek, 1870

placenta = flat cake + *ceras* = horn

The genus *Placenticeras* is characterized by a tightly coiled (moderately involute), lenticular (lens) shaped shell. The umbilicus is well rounded and moderately deep. The flanks are broad and convex and sometimes ornamented with ribs and tubercles along the umbilicus and midflank. The keel or venter is narrow and flat to slightly concave and protects the siphuncle which is near the surface of the shell. The suture pattern on *Placenticeras* is very complex with 10 to 14 divided, deep, narrow-necked saddles, lobes, and sinuses. *Placenticeras* has been recorded from less than 2 cm up to 90 cm in diameter. The genus has a world-wide distribution, and specimens have been recovered from South Africa, India, Siberia, Europe, and North America. Their remains are abundant within the Western Interior of North America from New Mexico to Alberta.

Placenticeras sancarlosense Hyatt, 1903

This species attains a moderate size for the genus of 24 cm in diameter. The umbilicus is moderately evolute; the venter is rounded and wide. The umbilical wall is low but steep in the juvenile stage and sloped in the mature adult. The umbilical shoulder has sparse, broad tubercles that radiate low, broad ribs. A distinguishing feature of *Placenticeras sancarlosense* is very prominent, ventrolateral tubercles that are evenly spaced on the ends of the primary and secondary ribs. This species has been found in the uppermost part of the Mancos Shale and the basal part of the Mesaverde Formation in the upper Rio Grande Valley of New Mexico.

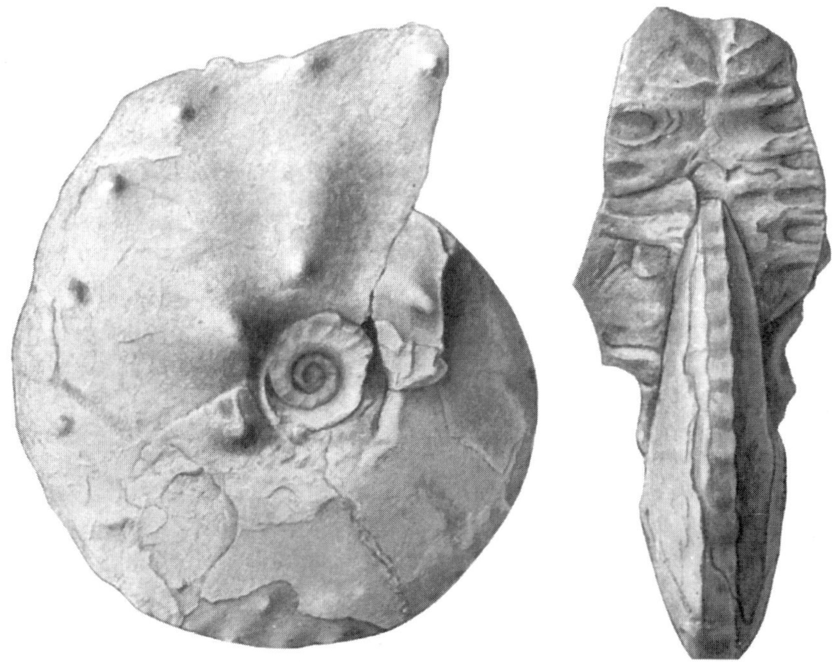

Placenticeras sancarlosense
USNM 73379
Manco: Santa Fe Co., NM
15 cm high

Placenticeras guadalupae (Roemer, 1852)

Placenticeras guadalupae has a thick, stout shell at all stages of growth with a moderately evolute umbilicus. It attains a fairly large size for the genus of up to 32 cm in diameter. The species is ornamented with several rows of nodes, one above the umbilical margin, one near the ventral edge, and finally, a row of small clavate tubercles along either side of the venter. The species becomes almost rectangular to circular in the whorl section of the adult form. Ribs are broad and faint. *Placenticeras guadalupae* has been found in Texas, in the upper part of the Mancos Shale of east central Utah, and in the upper Rio Grande Valley of New Mexico.

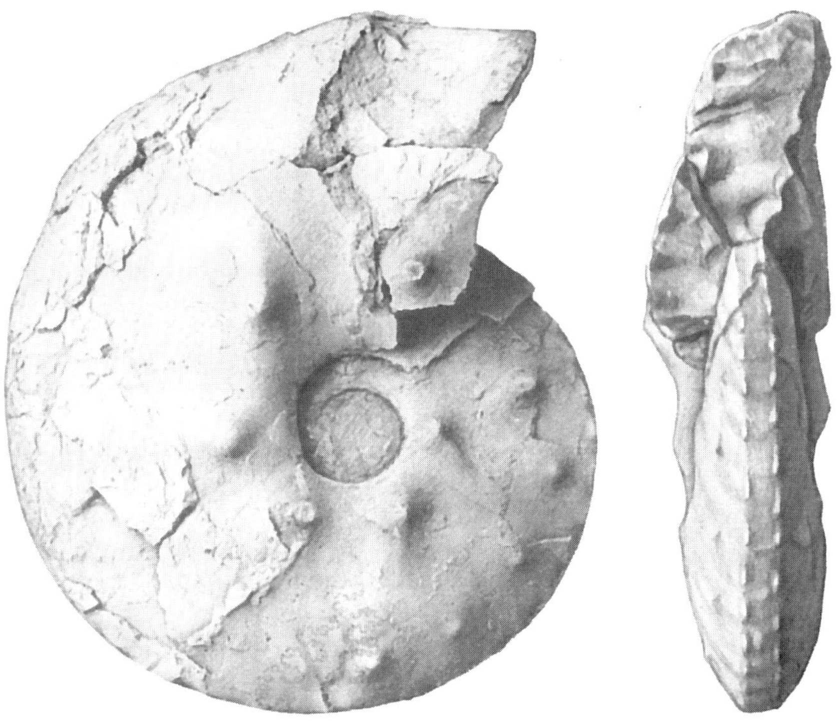

Placenticeras guadalupe
USNM 73381
14.5 cm across
Left: Side view Right: Venter view

Placenticeras planum Hyatt, 1903

Placenticeras planum is a species of medium size with the largest known specimens being no more than 25 cm in diameter. The most distinguishing feature is its smooth venter. In early stages, the venter is at first slightly concave, it then becomes flat, and finally rounds. *Placenticeras planum* has a double convex near-circular-shaped shell with a moderately involute umbilicus. It also may have tubercles along the umbilical shoulder, although some forms are smooth. The suture pattern is simpler than that of *Placenticeras meeki* or *Placenticeras intercalare*. The suture pattern increases in size to the third lateral lobe from the ventral side and then decreases beyond it. *Placenticeras planum* is found in the lower part of the Pierre Shale on the western rim of the Black Hills, in the Eagle Sandstone and Telegraph Creek Formation of Montana and Wyoming, in the Mancos Shale of central Utah, and in the uppermost Mancos Shale and lowermost Mesaverde Formation in northern New Mexico.

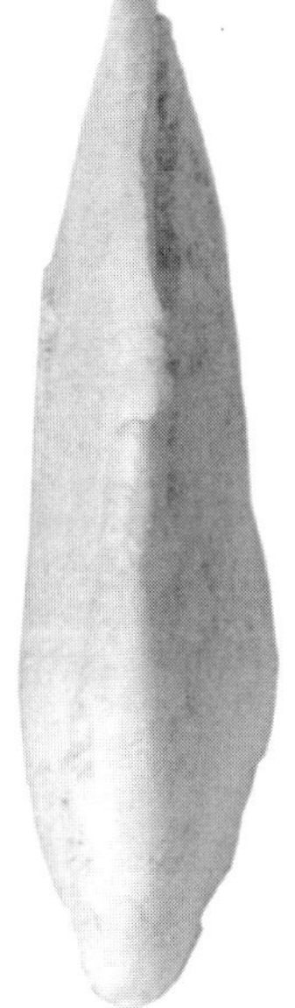

Placenticeras planum
Cast of USGS specimen
20.3 cm across

Placenticeras pingue Kennedy, Cobban, and Landman, 1996

This species displays a thick whorl section, nearly rounded venter, and a moderately evolute umbilicus. The name *pingue* means "fat" in Latin. This species name refers to the inflated or thick flanks. This form is similar to a later species, *Placenticeras intercalare*, with umbilical shoulder tubercles that each give rise to one or two low, prorsiradicate ribs. The ribs in turn form broad bullate-shaped tubercles that end about three-quarters of the distance from the umbilicus to the ventral edge. The venteral edge is flat early to rounded as it matures, and low clavi tend to alternate on either side of the venter. The suture pattern is significantly more simple than that of its descendant *Placenticeras intercalare*. There are 11 to 12 low lobes and saddles. The size appears not to exceed 27 cm in diameter. The species has been found from the *Baculites gregoryensis* and the *Baculites scotti* Range Zone of South Dakota and Wyoming.

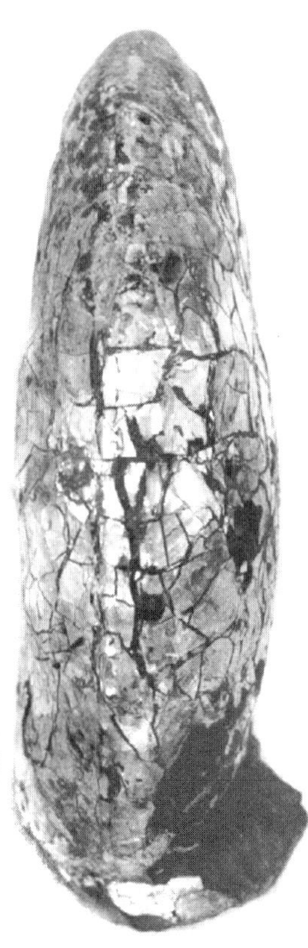

Placenticeras pingue
Above: BHI 4644
19 x 22 cm
Right: Ventral view
BHI 4646
19 cm high

Placenticeras meeki Böhm, 1898

Placenticeras meeki follows closely the description of the genus. The distinguishing features are its smooth convex, broad, nearly unornamented flanks and smooth venter. Some specimens occasionally possess small tubercles along the umbilical shoulder and some slight, broad, low ribbing. *Placenticeras meeki* has a large size range up to 80 cm in diameter. The narrow, concave venter in the smaller specimens becomes more rounded in larger shells. The species occurs in the Pierre Shale, sparsely in the *Exiteloceras jenneyi* Zone, and abundantly in the *Didymoceras cheyennense* through *Baculites cuneatus* Range Zones throughout the Western Interior.

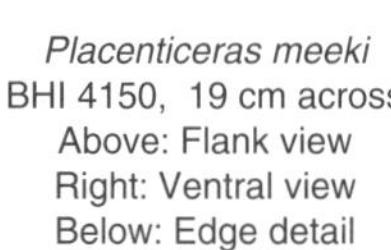

Placenticeras meeki
BHI 4150, 19 cm across
Above: Flank view
Right: Ventral view
Below: Edge detail

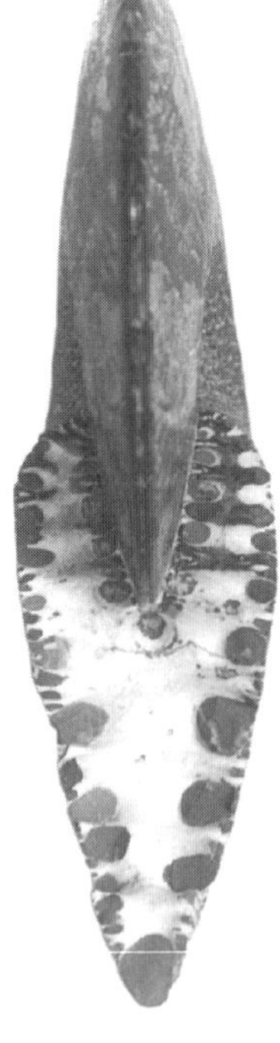

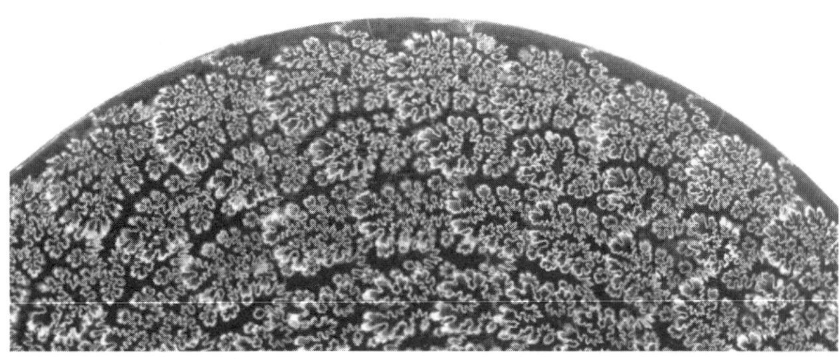

Placenticeras intercalare Meek and Hayden, 1860

The obvious differences between this species and *Placenticeras meeki* are its smaller size (up to 25 cm), thicker whorl section, narrower flanks, and greater ornamentation. There are two rows of well-defined tubercles on the flanks, one near the umbilicus and the second approximately two-thirds of the distance from the umbilicus toward the ventral edge. Minor broad, low, rounded ribs seem to connect these tubercles. The venter is slightly wider than that of the *Placenticeras meeki* and has small clavate tubercles on both edges. *Placenticeras intercalare* is distinguishable by having on the average three clavi along the ventral edge between each pair of outer flank tubercles. It is difficult to distinguish between macroconchs or microconchs in this species, although the macroconchs seem to have more rounded, thicker flanks, whereas the microconchs seem to be more depressed in the flanks. *Placenticeras intercalare* is found primarily in the *Exiteloceras jenneyi* through *Baculites compressus* Range Zones from New Mexico north into Canada.

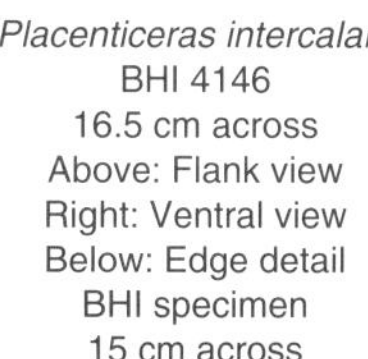

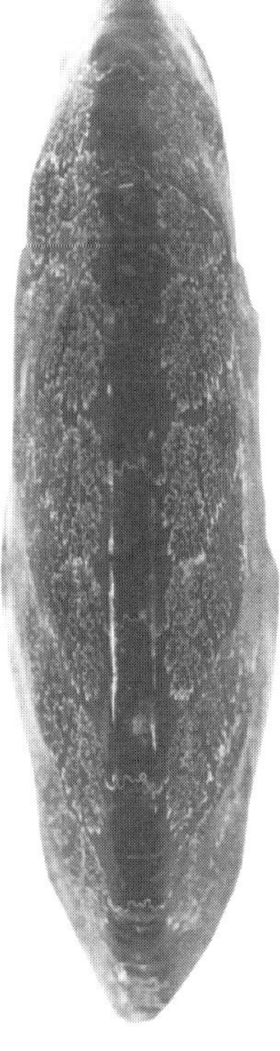

Placenticeras intercalare
BHI 4146
16.5 cm across
Above: Flank view
Right: Ventral view
Below: Edge detail
BHI specimen
15 cm across

Placenticeras costatum Hyatt, 1903

This species represents the last reported occurrence of Placenticeras within the Western Interior. As with its predecessor, *Placenticeras intercalare*, this form has a thick whorl section, broad flanks, and prominent tubercles. There are generally two rows of tubercles, one row above the umbilical shoulder and the other about two-thirds the distance from the umbilicus to the ventral edge. This form differs from its predecessor both in the suture pattern and in its ventral edge. The three outer suture lobes are more constricted than those of *Placenticeras intercalare*. The clavi, along either side of the ventral edge, number five between each pair of tubercles on the outer flank. The ribs are low, broad, and prorsiradiate, originating on the umbilical tubercles and splitting into a pair of ribs that develop into the conical-shaped tubercles furthest out on the flank. The species has been reported from the *Baculites cuneatus* and *Baculites reesidei* Range Zones of Colorado, South Dakota, Montana, Cypress Hills of Saskatchewan and possibly Alberta, and also in the *Baculites reesidei* Range Zone of New Mexico.

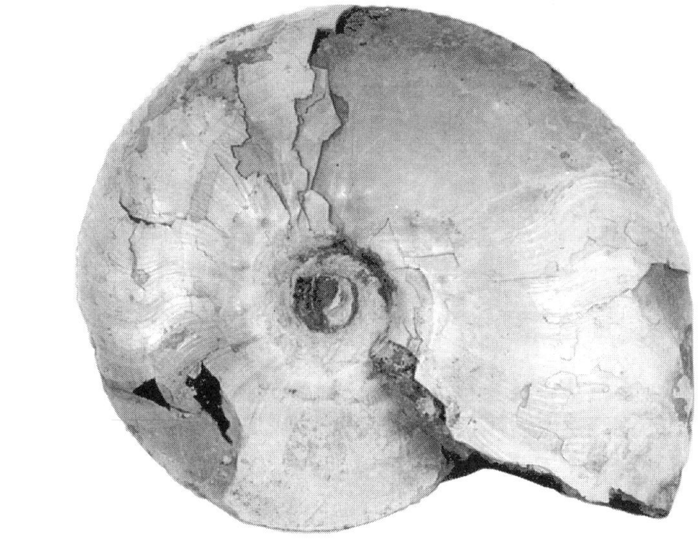

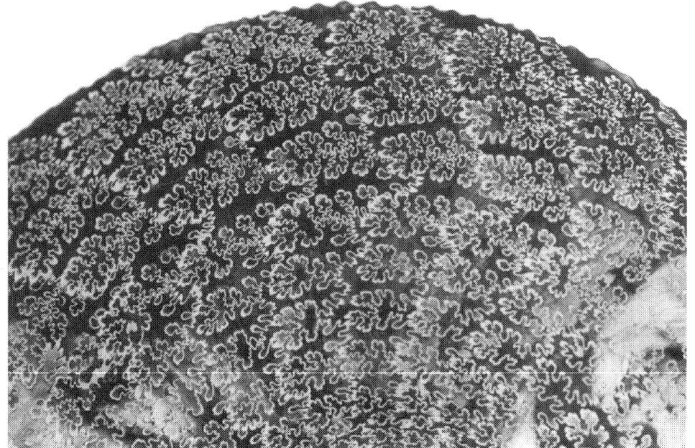

Placenticeras costatum
BHI 4519
17 x 20 cm
Above left: Side view
Above right: Edge view
Left: Close-up view
BHI 4520

Family Scaphitidae Meek, 1876

The Family Scaphitidae is characterized by its unique J-shaped body chamber. Having descended from an Albian and Turonian lineage, scaphites became more abundant and varied throughout the Western Interior during the early Pierre Shale (Campanian). Conch size can vary from extremely small (less than 2 cm) to quite large (greater than 15 cm in length). The shell can be compressed to inflated with a tightly coiled (involute) phragmocone; the body chamber is not curved over the phragmocone but rather drops straight down until curving back toward the phragmocone at the aperture. Lateral ribs are generally straight and branching. The ventrolateral ribs are also straight. Commonly, there are tubercles or bullae located along the umbilical shoulder and ventrolateral tubercles located on the body chamber. The family appears to be mostly limited to the Albian through Maastrichtian Cretaceous marine sediments of the northern hemisphere.

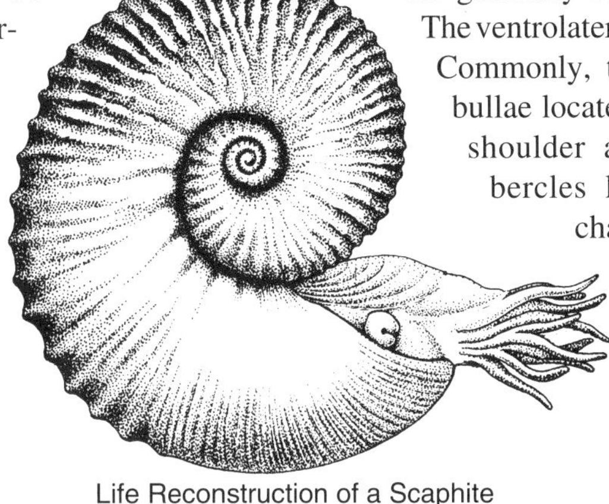

Life Reconstruction of a Scaphite
Illustration by D.S. Norton

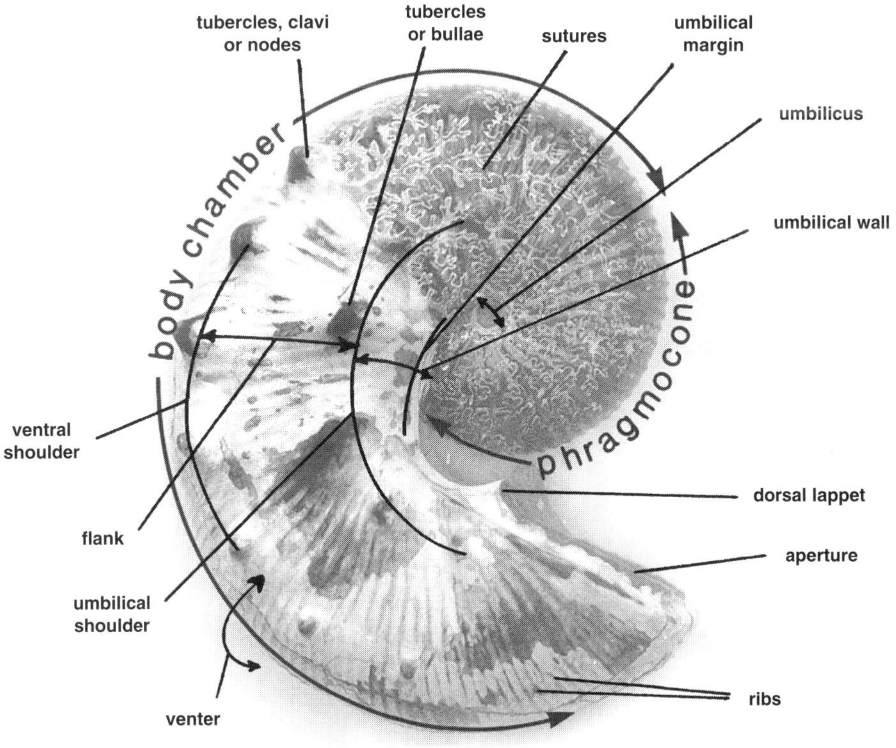

Shell Ornamentation
Family Scaphitidae
Illustration by N. Larson

Genus *Haresiceras* Reeside, 1927
after C. J. Hares + ceras = horn

The distinguishing features of this ammonite are a very involute phragmocone and tightly coiled body chamber. The whorl section is compressed and the venter is flat with ventrolateral nodes. The ribs are slightly flexuous, fine, and widely spaced, with a forward bend. The suture pattern is complex and shows a typical scaphite pattern in the large size and form of the first lateral saddle. The three species of *Haresiceras* from the Pierre Seaway occur in the Gammon Ferruginous Member of the Pierre Shale on the northern flank of the Black Hills. *Haresiceras placentiforme* Reeside, 1927, also occurs in the Telegraph Creek Formation in northwestern Wyoming and Cody Shale on the southeastern part of the Big Horn Basin. *Haresiceras natronense* Reeside, 1927, has also been found in the Cody Shale of western and central Wyoming and in the upper part of the Mancos Shale in northwestern Colorado. The genus occurs within the *Scaphites hippocrepis* Range Zones, and is limited to the Western Interior region of North America.

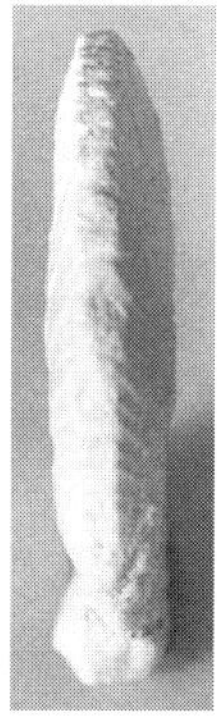

Left: Side and ventral view
Haresiceras fisheri
cast of USNM 73387
2.7 cm high

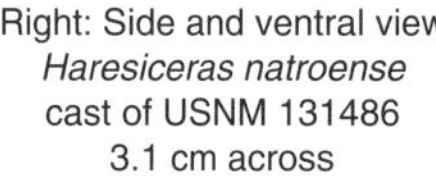

Right: Side and ventral view
Haresiceras natroense
cast of USNM 131486
3.1 cm across

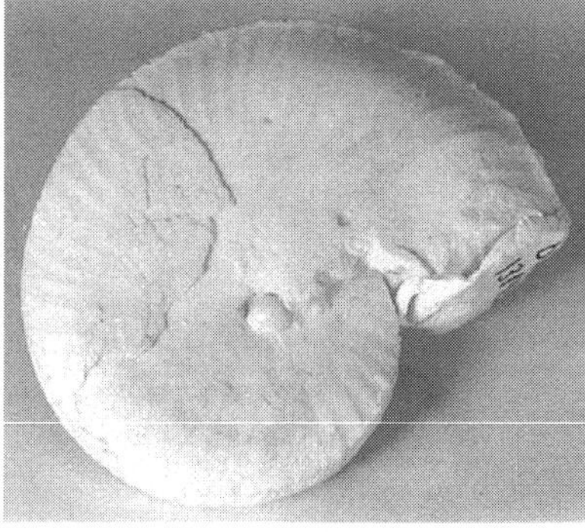

Left: Side and edge view
Haresiceras placentiforme
cast of USNM 131483
2.6 cm across

Genus *Hoploscaphites* Nowak, 1911
hoplo = heavily armed foot soldier + *skaphe* = boat + *ites* = a stone

Hoploscaphites is small to medium sized for the family with compressed or flattened shells, tightly coiled (involute) phragmocone, shallow umbilical walls, a short, slightly curved body chamber, and a slightly recurved hook. Ribs tend to be thicker on the phragmocone and become more dense and fine on the body chamber, often bifurcating and bending back (rursiradiate) toward the phragmocone at midflank. Small ventrolateral tubercles are often present on the body chamber and may or may not be present along the umbilical margin. *Hoploscaphites* seems to have its origin in the *Baculites obtusus* Zone of Greenland and continues through the *Jeletzkytes nebrascensis* Zone of the Fox Hills Formation. *Hoploscaphites* is present in the Campanian and Maastrichtian of North America and Europe.

Hoploscaphites gilli (Cobban and Jeletzky, 1965)

Hoploscaphites gilli
Left: BHI 4116 (macroconch) 5 cm high
Right: BHI 4117 (microconch) 4.5 cm high

Hoploscaphites gilli is a small- to medium-sized scaphitid (up to $4^1/_2$ cm) that occurs in the *Baculites perplexus* through the *Didymoceras nebrascense* Range Zones within the Pierre Shale. The shell is compressed (having a narrow whorl section), and the body chamber is broad and large and loosely coiled around the phragmocone in the macroconch. The microconch has a longer, straighter body chamber that separates the phragmocone from the recurved hook. The umbilical walls are tapered and shallow, flanks are flat to semirounded, and the venter is very rounded and blends into the flanks. The ribs are more coarse on the phragmocone but become very fine and dense on the body chamber. Ventrolateral tubercles may be present but are usually absent. Macroconchs of the species tend to have a much wider flank and an umbilical swelling on the body chamber and are about 25 percent larger than the microconchs.

Hoploscaphites landesi Riccardi, 1983

This species has a relatively small shell (2–3 cm). It is involute on the phragmocone, with the body chamber extending slightly beyond the phragmocone, leaving only a slight gap near the

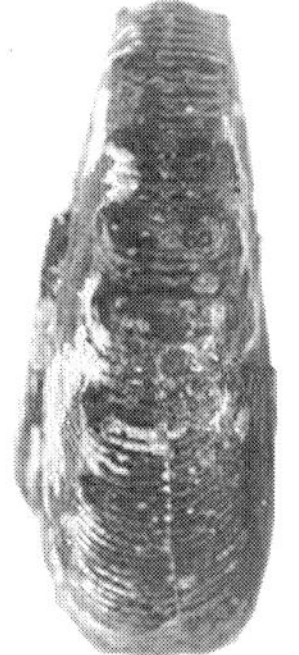

Hoploscaphites landesi
Left: BHI 4118 (macroconch) 4.2 cm high
Right: BHI 4119 (microconch) 3.5 cm high

Left: BHI 4118 (macroconch)
Right: BHI 4119 (microconch)

aperture. Ribbing is coarse on the phragmocone and becomes finer and more dense on the body chamber. There are slight bullae or tubercles along the umbilical margin and on the ventrolateral shoulder of the last part of the phragmocone and first half of the body chamber. The macroconch is distinguished by its slightly larger size, wider flanks, and a slight swelling above the umbilicus on the early part of the body chamber. *Hoploscaphites landesi* has been found throughout the Western Interior Seaway in the upper part of the *Didymoceras cheyennense* Zone through the *Baculites cuneatus* Zone.

Hoploscaphites melloi Landman and Waage, 1993

This is a moderately sized shell with fine dense ribs on all or most of the body chamber. Its ventral ribs are also very fine and dense. Ribs tend to curve away from the body chamber along the umbilical margin, bifurcate, and bend back toward the body chamber, then at two-thirds the distance across the flank, they bend back toward the phragmocone. This species is very compressed and similar to *Hoploscaphites nicolletii* from the Fox Hills Formation. There are few or no ventrolateral tubercles and none around the umbilical margin. Microconchs are commonly one-half to two-thirds the size of the macroconchs and have more prominent ornamentation and a more evolute umbilicus. The species has been reported in the upper part of the Mobridge Member of the Pierre Shale in north central South Dakota.

Hoploscaphites melloi (macroconch)
BHI 4120
4 cm high

Hoploscaphites nicolletii (Morton, 1842) **var.** *saltgrassensis* (Elias, 1933)

Elias figured and described this variety from phragmocones and other fragments that hc collected from the Pierre Shale in Wallace County, Kansas, along with another scaphite, *Jeletzkytes plenus*. Recently identified specimens from the Pierre Shale in Dawson County, Montana, not only extend their range, but also give us some complete specimens for description. This is a moderately sized (up to 8 cm), compressed, finely ribbed conch. The phragmacone tends to have prominent, straight primary ribs, with secondary ribs between them. The body chamber is slightly more inflated than the phragmocone with a strong recurved hook and prominent ventrolateral tubercles for nearly the entire length. The ribs of the body chamber are fine and become more dense toward the aperture.

Hoploscaphites nicolletii var. *saltgrassensis*
(macroconch)
BHI 4705
8.5 x 7.5 cm
Photo courtesy of Dr. Neil Landman

Hoploscaphites birkelundi Landman and Waage, 1993

The shell is very compressed with a strong recurved hook. The ribs are broad and low on the phragmocone, becoming narrower and more dense on the body chamber. Ventrolateral tubercles start on the body chamber near the phragmocone and usually quit when the body chamber curves back toward the phragmocone. Some macroconchs and most microconchs have umbilical bullae. As with all *Hoploscaphites*, the umbilicus is very involute. This species occurs with *Jeletzkytes dorfi*, but to date has only been found in the Fox Hills Formation of Niobrara County, Wyoming. The species should, however, occur throughout the Upper Pierre Shale above the *Baculites clinolobatus* Range Zone.

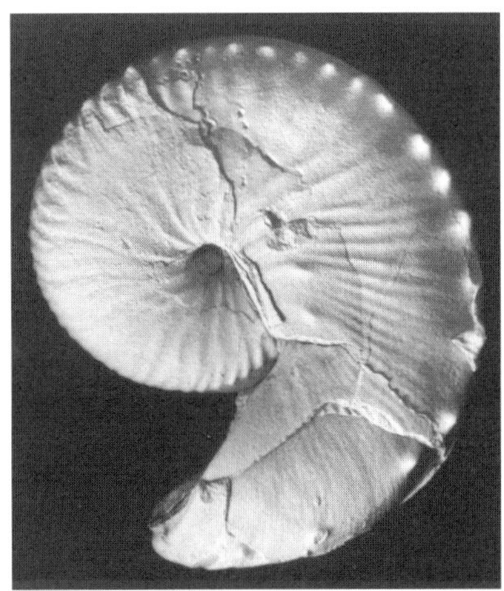

Hoploscaphites birkelundi
(macroconch)
Left: YPM 27170
6 x 4.7 cm
Right: YPM 27177
6.3 x 5.5 cm
Photos courtesy of
Dr. Neil Landman

Genus *Jeletzkytes* Riccardi, 1983
after J. A. Jeletzky + ites = a stone

Jeletzkytes can occur as small to large, moderately inflated conchs. The phragmocone is involute and the body chamber has a short shaft that extends slightly beyond the phragmocone with a weak recurved hook. The ribs are sparse and straight on the phragmocone but tend to bifurcate and bend slightly on the body chamber. There are prominent ventrolateral nodes beginning on the phragmocone that fade away near the aperture on the body chamber. Most species also have at least another row of tubercles around the umbilical margin or extending midway on the flank. Some species may also have two to three additional rows of tubercles on the flank starting on the phragmocone and fading away on the body chamber. The genus *Jeletzkytes* seems to first appear in the *Baculites obtusus* Zone and is present throughout the Maastrichtian, Fox Hills Formation, the last of the Cretaceous marine sediments in the Western Interior. It is found in Campanian and Maastrichtian marine sediments in Europe and North America.

Jeletzkytes nodosus (Owen, 1852)

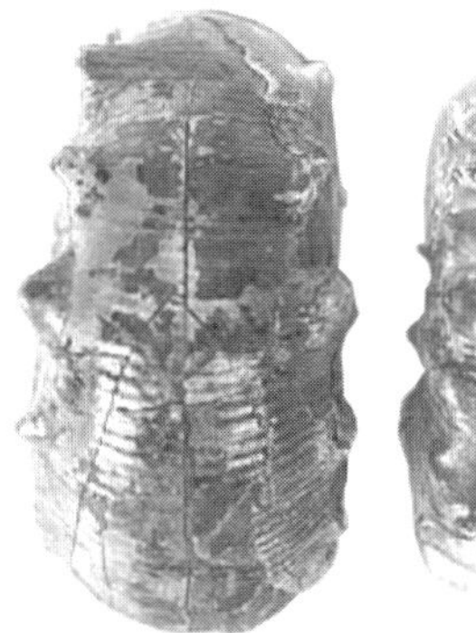

Jeletzkytes nodosus
Left: BHI 4121 (macroconch) 9 cm high
Right: BHI 4122 (microconch) 7 cm high

Ventral view
Left: BHI 4121
Right: BHI 4122

Jeletzkytes nodosus is comparatively large for this family (up to 11 cm). This species is quite robust with a somewhat long shaft on the body chamber and a moderate recurved aperture or hook. The umbilical walls are well rounded to steep. Ribs are broad, prominent, and fairly straight with some bifurcation and bending. The most distinguishing characteristic for *Jeletzkytes nodosus* in microconch and macroconch forms is its narrow, slightly rounded flank and a broad, rounded venter. The whorl section is more robust than any other scaphitids except for *Jeletzkytes crassus* and *Rhaeboceras subglobosum* from the Pierre Seaway. *Jeletzkytes nodosus* is further distinguished from other species by a prominent row of tubercles along the ventrolateral shoulder and another near the middle to the dorsal edge of the flanks. The venter width and the flank width at midbody chamber are of nearly equal size. Microconchs and macroconchs differ primarily in the size and robustness of the body chamber, with the microconchs somewhat compressed or narrower along the venter. *Jeletzkytes nodosus* ranges from the *Didymoceras cheyennense* through the *Baculites cuneatus* Range Zones. It has been found throughout the Western Interior Seaway and the Upper Campanian of the Gulf Coast.

Jeletzkytes brevis (Meek, 1876)

Jeletzkytes brevis can have small to large shells (up to 10 cm) that are somewhat compressed to well rounded and inflated. The body chamber has a short shaft and a recurved hook that extends slightly below the phragmocone. Ribs are thicker on the more robust forms and finer and sharper on the more compressed varieties. The ribs tend to bifurcate but remain fairly straight and prominent. The ventrolateral shoulder has a prominent row of tubercles that begins on the phragmocone and can extend the entire distance of the body chamber. A second row of tubercles lies on the flank above the rounded umbilical shoulder to one-third of the distance on the flank. The flanks of *Jeletzkytes brevis* tend to be only slightly rounded and quite wide compared to the narrow flanks of *Jeletzkytes nodosus*. *Jeletzkytes brevis* also is more compressed than *Jeletzkytes nodosus* and has a narrower and a more flattened venter. The flank width is about $1^1/_2$ times the diameter of the venter at midbody chamber. Microconchs tend to be about one-half to two-thirds the size of the macroconchs and much more compressed. The species is found from the *Didymoceras cheyennense* Range Zone through the *Baculites reesidei* Range Zone. *Jeletzkytes brevis* is common from Colorado through Alberta.

Jeletzkytes brevis (2 views)
Left: F47 BHI 4124 (macroconch) 8.5 cm high
Right: BHI 4123 (microconch) 6.6 cm high

Jeletzkytes "quadrangularis" (Meek and Hayden, 1860)
(invalid species)

This variety of *Jeletzkytes* has a small- to medium-size range, a nearly flattened venter, and somewhat compressed whorl section. Riccardi (1983) assigned two of the paratypes and the holotype of this species to the microconch of *Jelezkytes* cf. *brevis*. However, the holotype appears to be a microconch of *Jeletzkytes criptonodosus* and one paratype appears to be a microconch of *Jeletzkytes plenus*. The paratype of *Jeletzkytes plenus,* from the Yellowstone River area, does appear to be a microconch of *Jeletzkytes brevis*. The umbilical walls are narrow and steep, and there is a row of nodes directly above the walls on the umbilical shoulders. There is another row of tubercles on the ventrolateral shoulder that usually starts where the body chamber begins. This "species" has been described from microconchs only of other known species and should not be considered a distinct species. *Jeletzkytes "quadrangularis"* occurs with *Jeletzkytes nodosus* and *Jeletzkytes brevis* in the *Didymoceras cheyennense* Range Zone through the *Baculites cuneatus* Range Zone.

Side and Ventral view
Jeletzkytes quadrangularis paratype
USNM 365
(actually *Jeletzkytes plenus*)
6.8 cm high

Side and Ventral view
Jeletzkytes quadrangularis paratype
USNM 366
(actually *Jeletzkytes brevis* microconch)
5.5 cm high

Ventral and Side view
Jeletzkytes quadrangularis holotype
USNM 366
(actually *Jeletzkytes criptonodosus* microconch)
5.5 cm high

Jeletzkytes "furnivali" Riccardi, 1983
(doubtful species)

Jeletzkytes "furnivali" has flexuous ribs and a compressed whorl section that is subrounded to subrectangular. The holotype of the species appears to be a microconch, but Riccardi did not assign a macroconch to the species. This makes assignment as a distinct species doubtful . The authors feel that *Jeletzkytes "furnivali"* is possibly a microconch for an as yet undescribed *Jeletzkytes* macroconch from the *Baculites reesidei* Range Zone. It has a short shaft extending just below the phragmocone and curving strongly back, leaving it slightly more open than *Jeletzkytes brevis*. The umbilical wall is curved and shows no swelling, flanks are flat and parallel, and the venter is flat on the phragmocone to rounded on the body chamber. Ribs tend to be thick and bifurcate on the phragmocone and on much of the body chamber, but become finer as they near the aperture. A ventrolateral row of tubercles runs along the shoulder from the last part of the phragmocone to nearly the end of the body chamber where the tubercles become smaller. There is a row of nodes or small bullae that begin nearly at midflank on the phragmocone and move in just above the dorsum at the aperture. *Riccardi* stated that *Jeletzkytes "furnivali"* occurred from the *Baculites compressus* through the *Baculites reesidei* Range Zones of the Pierre Shale and age-equivalent rocks.

Side and Ventral view
Jeletzkytes furnivali Holotype
GSC 67093
8.5 cm high

Jeletzkytes crassus (Coryell and Salmon, 1934)

This species of *Jeletzkytes* is the most robust and well rounded of all the species. It is also quite large (up to 12.5 cm). The body chamber has a short shaft that quickly bends back on the phragmocone. The venter is very wide and broad, almost as wide as the flank. The species has a steep umbilical wall and slightly rounded flanks. The ribs are bifurcated, prominent, and remain fairly straight. The ventrolateral shoulder has a row of small tubercles that arise early on the phragmocone and end at the aperture. There is also a row of nodes above the umbilical shoulder beginning early on the phragmocone but fading out on the body chamber. The macroconch has a large swelling near the umbilicus on the early portion of the shaft. Microconchs are quite wide across the venter with narrow, slightly rounded flanks, and

Side view
Jeletzkytes crassus
Left: BHI 4125 (macroconch) 9.7 cm high
Right: BHI 4126 (microconch) 6.3 cm high

Jeletzkytes crassus continued

are ornamented with large, prominent tubercles. They are only one-quarter to one-half the size of the macroconchs. The maximum width of the venter is so wide, it is even wider than the flank width at the midbody chamber. *Jeletzkytes crassus* occurs commonly in the *Baculites eliasi* and *Baculites baculus* Zones and equivalent rock units of the Pierre Shale from Kansas through Alberta and Saskatchewan.

Jeletzkytes crassus
Side and venter view
Left: (macroconch) 10.2 across x 10.5 cm high
Right: (microconch) 3.5 across x 6.6 cm high
Jim and Joyce Grier collection

Jeletzkytes plenus (Meek and Hayden, 1860)

Jeletzkytes plenus is a robust species characterized by its steep umbilical wall, well-rounded venter and flanks, and large size (up to 11 cm). This species has fairly straight, medium to fine ribs that tend to bifurcate about two-thirds of the way out on the flanks. It also has small tubercles on the ventrolateral shoulder quite early on the phragmocone that continue almost all of the way to the aperture. There is an inner row of small lateral nodes or bullae above the umbilical shoulder that begin early on the phragmocone, but fade away on the body chamber. Occasionally, a third row of small tubercles is located midflank on the body chamber. The microconch for this species is slightly compressed in whorl section, has flat to subrounded flanks, and is ornamented by small tubercles on

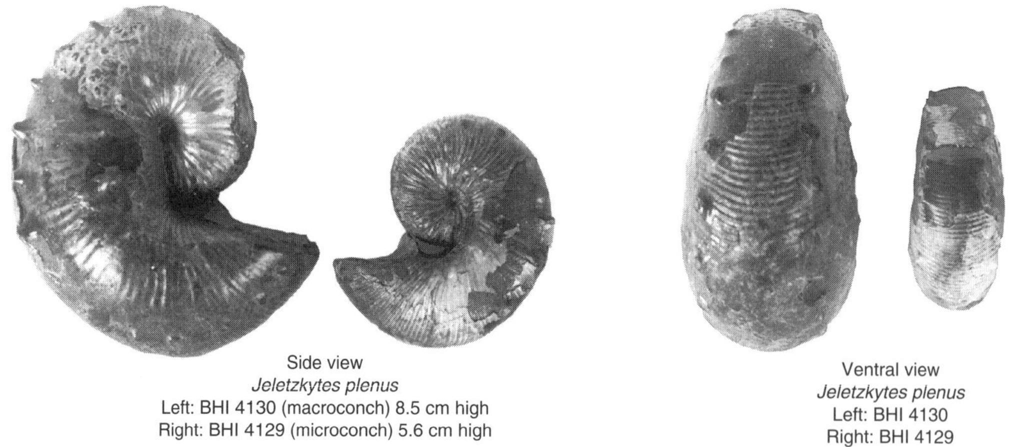

Side view
Jeletzkytes plenus
Left: BHI 4130 (macroconch) 8.5 cm high
Right: BHI 4129 (microconch) 5.6 cm high

Ventral view
Jeletzkytes plenus
Left: BHI 4130
Right: BHI 4129

Jeletzkytes plenus continued

Jeletzkytes plenus
(macroconch)
BHI 4128
10 cm across

the ventral shoulder and above the umbilical shoulder. Macroconchs are broader across the venter than *Jeletzkytes criptonodosus,* but less robust than *Jeletzkytes crassus.* The flank width is only slightly wider than that of the venter. The size ratio for microconch to macroconch is about one to two or one to three. *Jeletzkytes plenus* is found in the upper part of the Pierre Shale, primarily in the *Baculites eliasi* and *Baculites baculus* Range Zones.

Jeletzkytes criptonodosus Riccardi, 1983

This species is a medium-sized conch (up to 9 cm) similar to *Jeletzkytes brevis,* but differs by sparser, more irregular ribbing, more complex suture pattern, and the presence of tubercles on the innermost whorls of the phragmocone. The umbilical wall is steep to rounded with weakly curved to flat flanks and a broadly rounded venter. The lateral ribs are strong, rounded, and tend to pair from the lateral tubercles above the umbilicus toward the ventrolateral tubercles above the venter. Ribs tend to be very straight on the phragmocone to slightly curved on the body chamber. The venter to flank ratio follows that of *Jeletzkytes brevis,* being about 1 to 1$\frac{1}{2}$ at midbody chamber. Macroconchs

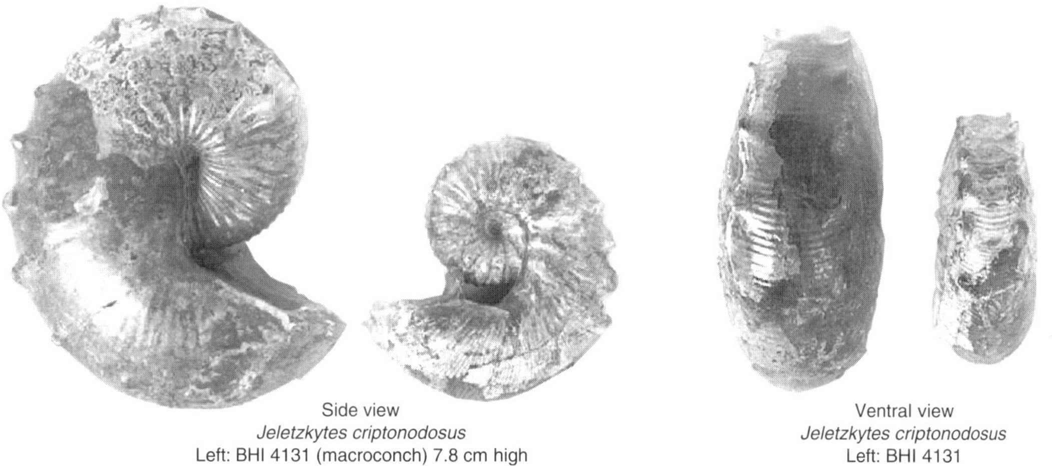

Side view
Jeletzkytes criptonodosus
Left: BHI 4131 (macroconch) 7.8 cm high
Right: BHI 4132 (microconch) 5.2 cm high

Ventral view
Jeletzkytes criptonodosus
Left: BHI 4131
Right: BHI 4132

Jeletzkytes criptonodosus continued

are large and compressed like *Jeletzkytes brevis,* whereas the microconchs are small and compressed with much broader ribbing and larger ventrolateral tubercles, and are one-half to one-third the size of the macroconchs. *Jeletzkytes criptonodosus* occurs primarily in the *Baculites baculus* Zone, but similar specimens have been found as low as the *Baculites cuneatus* Zone.

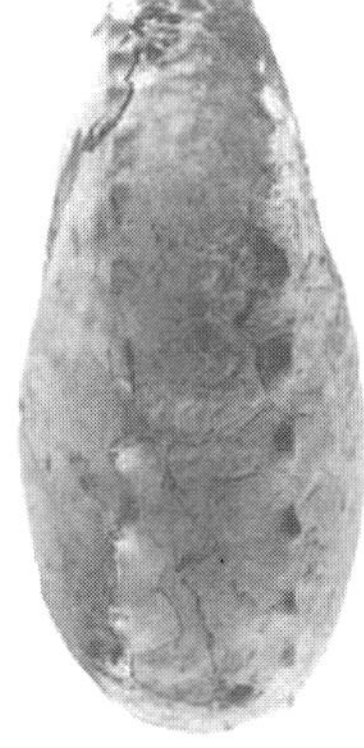

Type *Jeletzkytes criptonodosus* holotype
GSC 67104
8.6 cm high

Jeletzkytes dorfi Landman and Waage, 1993

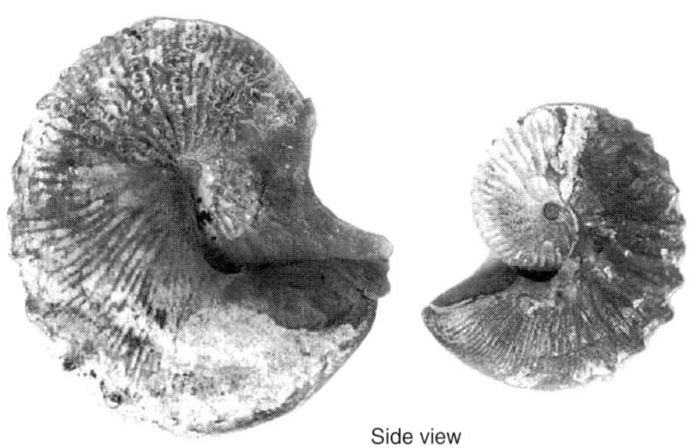

Side view
Jeletzkytes dorfi
Left: BHI 4133 (macroconch) 7.5 cm high
Right: BHI 4134 (microconch) 5.3 cm high

Ventral view
Left: BHI 4133
Right: BHI 4134

This species is of medium size for the genus (between 7 and 8.5 cm for macroconch size). The shell is somewhat compressed, with a long tapering body chamber and a high apertural angle on the recurved hook. The umbilical wall is short and steep with slightly rounded flanks and an almost flat venter. Ribs are strong and sparse on the phragmocone, becoming closer and finer near the aperture. It has many closely spaced ventrolateral tubercles, and there is another faint row of tubercles above the umbilical margin to one-third of the distance of the flank. The venter is rounded on the phragmocone to flat on the body chamber. The microconch for *Jeletzkytes dorfi* is about one-half the size of the macroconch, with broad, widely spaced ribs, a compressed whorl section, and prominent tubercles along the ventral shoulder with minor tubercles on the umbilical region. *Jeletzkytes dorfi* occurs above the *Baculites clinolobatus* Zone, in the *Jeletzkytes dorfi* Zone east of the Black Hills, and in the lowermost Fox Hills Formation, just west of the Black Hills in Wyoming.

Genus *Ponteixites* Warren, 1934

Ponteix = town in Saskatchewan + *ites* = a stone

Ponteixites appear to be the dwarfed Maastrichtian descendant of the Campanian *Rhaeboceras*. *Ponteixites* is quite small and compressed in whorl cross section. It has a slightly rounded, narrow umbilical wall, somewhat evolute, slightly rounded to flat flanks, and well-rounded yet narrow venter. The ribs are prominent, sparse, and well-rounded, with an average of two secondary ribs for every primary rib. The suture pattern is the same as for all scaphitids when the same-size specimens are compared. This genus is found only in the Maastrichtian of the Western Interior.

Ponteixites robustus Warren, 1934

Side and ventral view
Ponteixites robustus
BHI 3044
4 cm across

Of the two described species, *Ponteixites robustus* is the more robust, both in whorl section and in having broader, more prominent ribs on the flanks and on the venter. The umbilicus is moderately evolute (loosely coiled). *Ponteixites robustus* tends to have straight to slightly bent primary ribs and a secondary rib for each primary one beginning about the middle of the flank. The umbilical walls are shallow, and the flanks are slightly rounded. Complete mature *Ponteixites robustus* have a maximum diameter of 5.5 cm. *Ponteixites robustus* occurs in the *Baculites baculus* and *Baculites eliasi* Range Zones and is primarily found in Montana, Alberta, and Saskatchewan.

Ponteixites gracilis Warren, 1934

Ponteixites gracilis seems to be consistently smaller (1.4 cm to 2 cm in diameter), more compressed, and slightly more tightly coiled (involute) than *Ponteixites robustus*. The umbilical walls are also more shallow and the flanks contain a greater number of both primary and secondary ribs. The venter is well-rounded and has broad prominent ribs. This species may actually be the microconch for *Ponteixites robustus* since both occur in the same Range Zones and localities.

Side and ventral view
Ponteixites gracilis
BHI 628 2.2 cm across

Genus *Rhaeboceras* Meek, 1876

rhaibos = bent + *ceras* = horn

Rhaeboceras is a scaphitid of medium to large size with involute inner whorls and a more loosely coiled body chamber. The phragmocones tend to resemble finely ribbed *Pachydiscus*. The whorl section is rounded to slightly compressed. The umbilical walls are steep, the flanks are slightly rounded, and the venter is very well rounded. Straight to slightly curved, bifurcating ribs are present on the flanks, and the venter has broad straight ribs. Tubercles may or may not be present above the umbilical margin and above the venter of the body chamber. The development of tubercles may help determine the gender (macroconch, microconch) of the ammonites. The genus *Rhaeboceras* may be limited to the Western Interior of North America.

Rhaeboceras halli (Meek and Hayden, 1856)

Rhaeboceras halli is a fairly small (up to 15 cm) and compressed variety of the genus. Its ribs are fine and closely spaced; some bifurcate about halfway up the flank with several fine secondary ribs arising in between the straight ones within the middle third of the flank. Tubercles or clavi may occur above the umbilical margin near the end of the phragmocone and on the body chamber. This species occurs infrequently in the upper part of the Pierre Shale, but is most common in the *Baculites reesidei* and *Baculites jenseni* Range Zones of Montana.

Side and ventral view
Rhaeboceras halli
BHI 3044   14 cm across

Rhaeboceras albertense (Warren, 1930)

Rhaeboceras albertense is a medium-size conch for the genus (up to 18 cm), with an ovate to subrectangular whorl section and steep almost vertical umbilical walls. The whorl section near the end of the body chamber becomes nearly as wide as it is high. It differs from *Rhaeboceras halli* by its straighter, coarser, and broader ribs, and also by its very prominent secondary ribs that begin on the middle third of the flank and extend over the well-rounded venter. As in *Rhaeboceras halli*, tubercles or clavi may be present along the umbilical margin of the body chamber. The species occurs in the *Baculites reesidei* and *Baculites jenseni* Range Zones in the Pierre Shale and its equivalents, primarily in Montana, Alberta, and Saskatchewan.

Side and ventral view
Rhaeboceras albertense
BHI 4136
14 cm across

Rhaeboceras subglobosum (Whiteaves, 1885)

This species is the largest (up to 30 cm) and the most robust of the *Rhaeboceras*. The whorl section is almost round, the umbilical walls are very steep, and the flanks and venter are very rounded. Broad, straight, primary, and secondary ribs decorate the flanks and venter. There appear to be no tubercles or clavi. Its suture pattern is also the most complex of all the *Rhaeboceras*. The species occurs primarily in the *Baculites reesidei* and *Baculites jenseni* Range Zones of Montana, Alberta, and Saskatchewan.

Side and ventral view
Rhaeboceras subglobosum
BHI 4137 11 cm across

Rhaeboceras coloradoense Cobban, 1987

This species was assigned as the "noded" variant by Cobban when he wrote the description of the species. Out of all the species within this genus, this one has the most typical look of a scaphite. The species has a row of tubercles on the body chamber above the umbilical margin. It also has flattened flanks and a rather compressed cross section. The ribs are prorsiradiate, flexuous, high, rounded, and evenly spaced. There are also secondary ribs in between the primary ribs on the body chamber. The species has been found in the Pierre Shale in the *Baculites jenseni* Range Zone in Colorado, and the *Baculites eliasi* Range Zone of Dawson County, Montana.

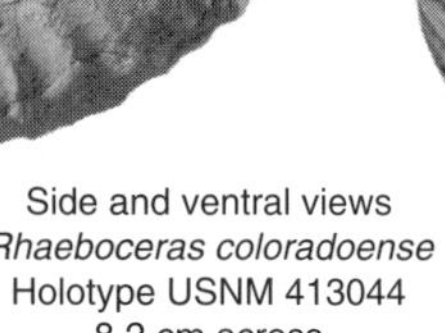

Side and ventral views
Rhaeboceras coloradoense
USNM 413046
4.5 cm across

Side and ventral views
Rhaeboceras coloradoense
Holotype USNM 413044
8.2 cm across

Rhaeboceras burkholderi Cobban, 1987

Rhaeboceras burkholderi is defined by a compressed cross section, flattened flanks, and a slightly flattened venter that is bordered by bullae. The ribbing is prorsiradiate and slightly flexuous, with the primary and secondary rigs giving rise to the bullate tubercles on the ventrolateral shoulder.

The species is somewhat smaller than the other species, though only slightly smaller than *Rhaeboceras coloradoense*. The species has been found in both the Pierre and Bearpaw Shales from the *Baculites reesidei* through the *Baculites eliasi* Range Zones in Saskatchewan and Montana.

Side and rear view
Rhaeboceras burkholderi holotype
USNM 413047
5.7 cm across

Rhaeboceras mullananum (Meek and Hayden, 1862)

As Cobban (1987) notes on *Rhaeboceras*, the species was collected by Meek and Hayden about 20 odd miles beneath Fort Benton, Montana, which places it in the Santonian Age, Marias River Shale. That would make this specimen the earliest *Rhaeboceras* discovered. The form appears to be much later than Santonian and should be from the Campanian rocks of Montana, but only one specimen of this species (the type specimen) has been found to date. The authors believe that it came from the Bearpaw Shale, so we felt obligated to include it in this publication.

Side and front view
Cast of holotype
Rhaeboceras mullananum
USNM 1924
7.2 cm across

Genus *Scaphites* Parkinson, 1811

skaphe = a boat + *ites* = a stone

This genus is described as being either compressed or robust. The phragmocone coiling is involute; the body chamber shaft may be either short or moderately long, but it is not curved over the phragmocone. The aperture is somewhat constricted and usually has a long dorsal lappet. Ribs are generally rectiradiate and branch or become intercalatory on the phragmocone. Tubercles on the umbilical and ventral shoulders are normally present on the body chamber. This genus has a worldwide distribution.

Scaphites hippocripis I
USNM 160280
Left: Macroconch
Above: Microconch

Scaphites hippocripis II
Left: Macroconch
Above: Microconch

Scaphites hippocripis III
Left: Macroconch
Above: Microconch

Scaphites hippocrepis (DeKay, 1827)

Scaphites hippocrepis is found in the uppermost part of the Niobrara and in the lower part of the Pierre Shale, in the Eagle Sandstone, in the Cody Shale, and in the Telegraph Creek Formation. Cobban (1969) assigned the Roman numerals I, II, III to designate three subspecies. They are scaphitids of a small to below average size (1–1^{1}/2 in.) with a very loosely coiled body chamber similar to the earlier Turonian Age scaphites from the Carlile Formation. All types have prominent ventral ribs and sparse lateral ribs. They have ventrolateral tubercles on the body chamber and some near the umbilicus. *Scaphites hippocrepis* I, from the Niobrara and lowermost Pierre Shale, has the fewest ribs and tubercles; *Scaphites hippocrepis* II of the Pierre Shale has more ribs and ventrolateral tubercles; and *Scaphites hippocrepis* III, also from the Pierre Shale, is even more densely ribbed, and has umbilical and midflank tubercles and ventrolateral tubercles that extend onto the phragmocone.

Genus *Trachyscaphites* Cobban and Scott, 1964
trachy = rough + *skaphe* = boat + *ites* = a stone

This genus has a very ornate conch of small to medium size. A multitude of tubercles, clavi, bullae, and ribs decorate this shell from early on the phragmocone through the body chamber. The phragmocone is very tightly coiled (involute), and the body chamber is straight and separates from the phragmocone before it bends back at the aperture. There are as many as five rows of tubercles on the flank of the phragmocone and body chamber that decrease in size from the venter toward the umbilicus. Ribs are straight, tend to join and go between the tubercles from both sides, and are spaced very tightly. This genus is similar to the scaphitid, *Discoscaphites gulosus*, of the Fox Hills Formation. *Trachyscaphites* is limited to the Campanian of North America and Europe.

Trachyscaphites pulcherrimus (Roemer, 1841)

Trachyscaphites pulcherrimus resembles the description of the genus very closely. The whorl section is compressed to rounded, and the umbilical walls are steep. There are five rows of tubercles on each flank; two rows of ventrolateral tubercles parallel each other, typically a row of tubercles above the umbilical margin and generally one to two more rows of tubercles between these sets. The tubercles get larger and even disappear on the body chamber, and ribs tend to join these tubercles in straight lines. *Trachyscaphites pulcherrimus* is more slender, slightly more involute, and has a less-extended body chamber than any other species of *Trachyscaphites*. *Trachyscaphites pulcherrimus* occurs in the *Baculites perplexus* and *Baculites gregoryensis* Range Zones of the Pierre Shale in Colorado, Wyoming, and South Dakota.

Trachyscaphites pulcherrimus
BHI 2148
3.5 cm across

Side and ventral view
Trachyscaphites pulcherrimus
Jim Schoon collection
4.1 cm high

Trachyscaphites redbirdensis Cobban and Scott, 1964

Trachyscaphites redbirdensis is a large species of up to 11 cm high by over 5 cm thick on the venter. It has a long straight shaft on the body chamber and a strongly retracted hook. There are five

rows of nodes on each flank, all almost equally spaced. All of the nodes are prominent, but the ventrolateral tubercles are the largest. The umbilical nodes are somewhat bullate in form. There is also a row of nodes on either side of the center of the venter. *Trachyscaphites redbirdensis* occurs in the Redbird Silty Member of Wyoming.

Side and rear view
Trachyscaphites redbirdensis
Holotype USNM 132309
9.6 cm across

Trachyscaphites spiniger (Schlüter, 1872) **subspecies** *porchi* (Adkins, 1929)

This species is a moderately large conch that differs from others of the genus by having four rows of nodes on each flank. There are also two rows of tubercles between the ventrolateral tubercles on the venter. The form is stout and robust, with a long separated body chamber and a strongly retracted aperture. The species has straight to slightly flexuous ribs that tend to join and separate the tubercles. The species has been reported from Germany, Russia, Sweden, and Belgium.

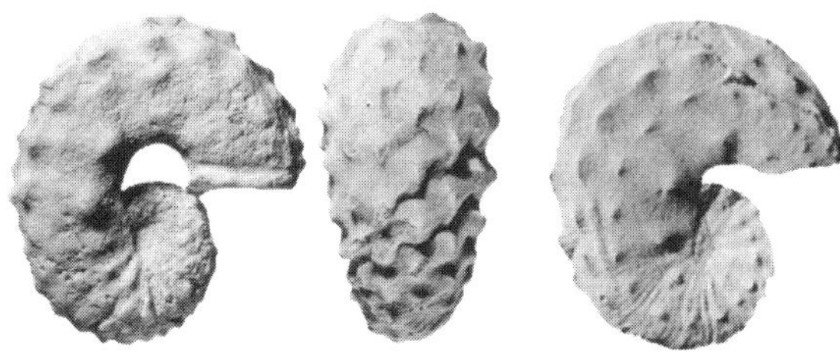

Trachyscaphites spiniger porchi described and figured here has been found in Texas, Montana, Colorado, and Kansas.

Trachyscaphites spiniger subspecies *porchi*
Left: Internal mold USNM 132320 4.4 cm across
Center: Rear view USNM 132319 4.7 cm high
Right: Internal mold USNM 132324 4.5 cm across

Trachyscaphites praespiniger Cobban and Scott, 1964

Trachyscaphites praespiniger differs from the other species of *Trachyscaphites* by having only three and rarely four rows of nodes on each flank. Generally only the midlateral row persists to the aperture on the body chamber. The species is moderate to large in size, ranging between 8.5 and 13 cm high. It is moderately robust on the ventral

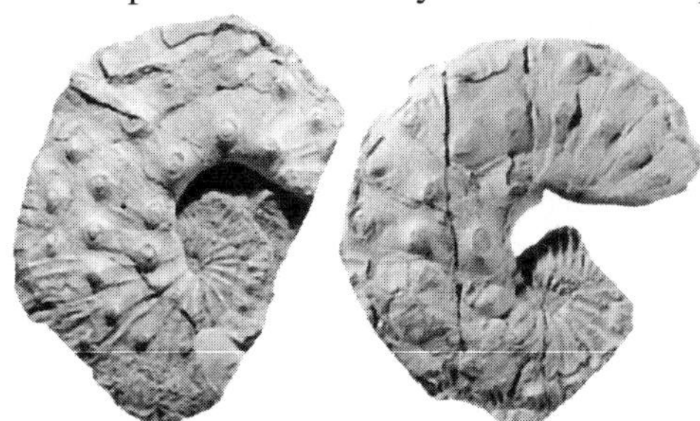

side. The body chamber does not separate from the whorls as much as on *Trachyscaphites redbirdensis* or *Trachyscaphites spiniger porchi*. The species has been reported from South Dakota, Montana, Wyoming, Colorado, and Utah.

Trachyscaphites praespiniger
Left: Holotype USNM 132333 7.6 cm high
Right: USNM 132335 7.2 high

This family consists of small to large species that tend to have very involute or closed umbilici. The shell is compressed, lenticular, and planispiral, with a sharp or narrowly rounded venter. Some genera possess weak lateral and ventrolateral tubercles, and faint ribs or swellings. Sphenodiscidae have been found on every continent except Antarctica, from the Upper Campanian through the Maastrichtian.

Genus *Sphenodiscus* Meek, 1876

spheno = a wedge + *discus* = a disk

The genus *Sphenodiscus* has been described as a smooth planispiral, lenticular-shaped shell with a closed or involute umbilicus. The flanks are wide and very broadly convex, and the venter is narrow and acute. Some species show low, broad, widely spaced ribs (or undulations) and occasional tubercles in early forms. The suture pattern is fairly simple, with about eleven or twelve rather short, deeply divided lobes. The genus attains a size of 50 cm in the Timber Lake Member of the Fox Hills Formation, but rarely gets larger than 15 cm in the Pierre Shale. Macroconchs are considerably larger than the microconchs, which tend to have prominent ribbing or undulations. The genus has been recorded from western Europe, the Middle East, India, Madagascar, Mexico, Venezuela, and the United States. A Gulf Coast species, called *Sphenodiscus pleurisepta* (Conrad, 1857), is occasionally found in the *Baculites clinolobatus* and possibly the *Jeletzkytes dorfi* Range Zones of the Pierre Shale.

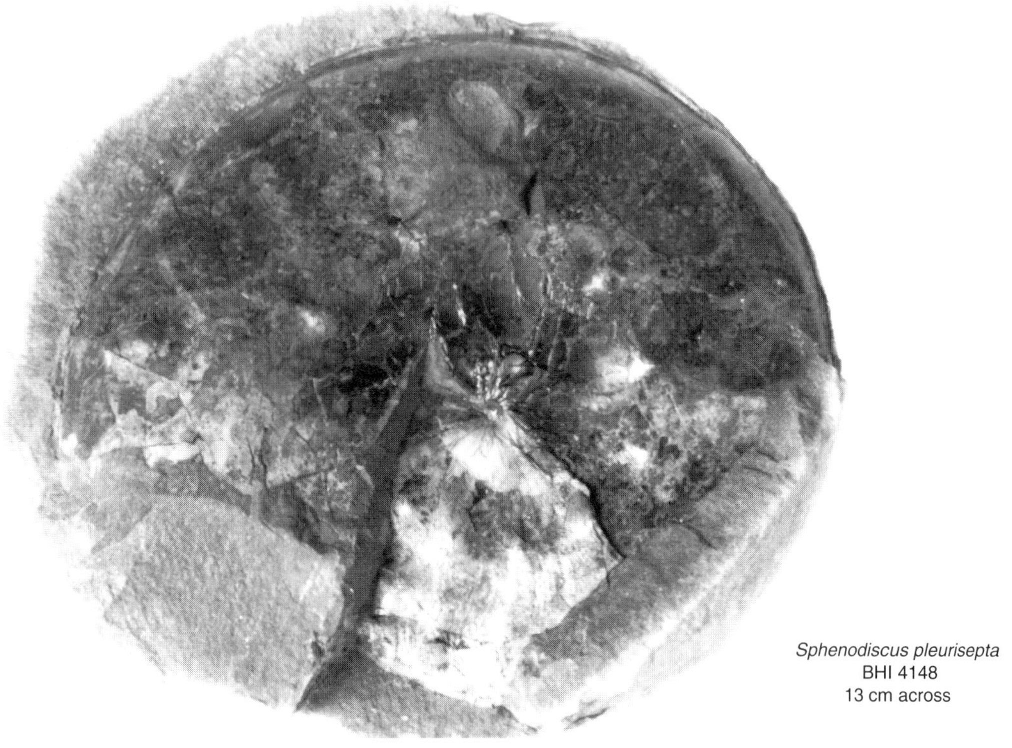

Sphenodiscus pleurisepta
BHI 4148
13 cm across

Genus *Coahuilites* Bose, 1927
Coahuila - A state of Mexico

The shell of *Coahuilites* is disc-shaped with an involute umbilicus, rounded to irregular flanks, and two distinct rows of tubercles. One row of tubercles is located on the midflank and the other is on the ventral shoulder. The venter is characterized by a sharp keel early that changes to almost rooflike, and then becomes nearly flat in mature individuals. The genus has been found from northern Mexico to Redbird, Wyoming.

Coahuilites sheltoni Bose, 1927

This species of *Coahuilites* is the most abundantly found within the genus. The suture pattern seems to be the most distnguishing feature between this and the other described species of *Coahuilites*. Because it is the only species that has been found within the Western Interior, we will not explain those differences. The species portrayal follows closely the explanation of the genus. *Coahuilites sheltoni* has been found in northern Mexico (Coahuila), Texas, Colorado (?), and within the *Baculites clinobatus* Range Zone of Wyoming.

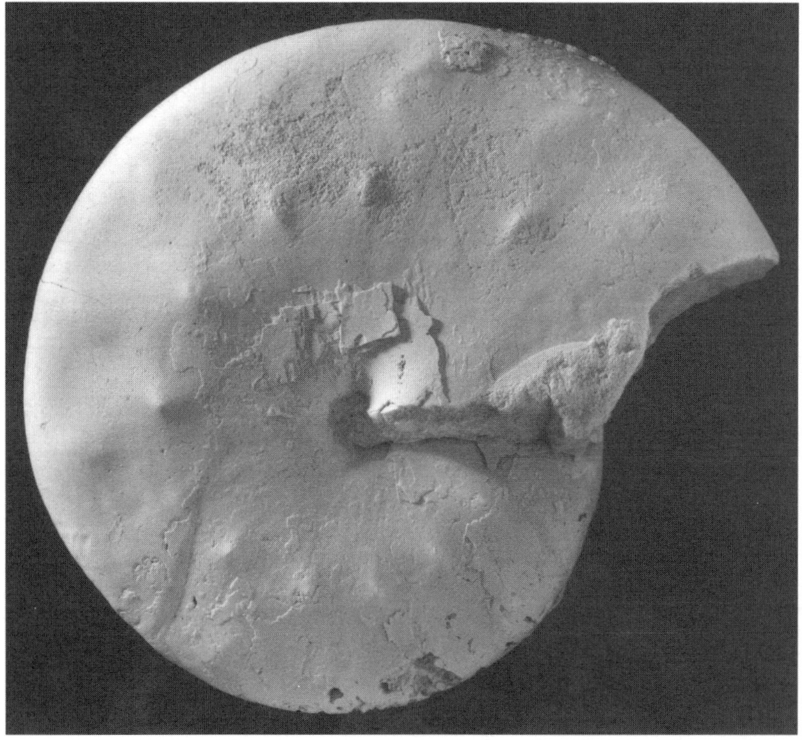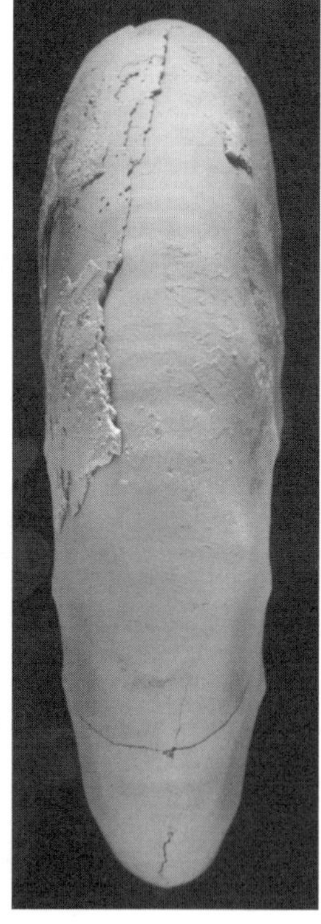

Coahuilites sheltoni
USMN 32152
Left: Side view 14.9 cm high x 17.1 cm across
Right: Ventral view
Photo courtesy of Dr. Neil Landman

ORDER BELEMNITIDEA

B elemnites were closely related to the modern-day squid, but had an elaborate internal shell including a phragmocone that resembles that of straight, externally shelled nautiloids of the Paleozoic Epoch. These animals were probably rapid swimmers that looked and behaved much like

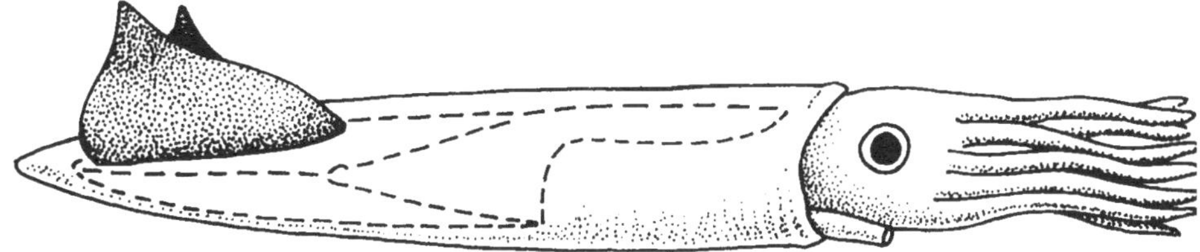

A Life Reconstruction of a Belemnite
[dashed line indicates hard internal shell - see illustration below]
Illustrations by D.S. No`rton

extant squids. Their bodies were shaped like a torpedo with two lateral fins that could steer the belemnite as it swam. Like the squid, it would propel itself through the water by taking water into its mantle cavity and force it out through an organ called a funnel. All representatives in this order had ten equal or subequal arms with hooklets. The belemnites' hard, internal shell consisted of a guard or rostrum, a phragmocone, a siphuncle, and a proostracum. The rostrum and proostracum were composed of concentrically layered calcite that radiates from the ventral surface. The rostrum encompassed the phragmocone and the proostracum provided a hard dorsal protection for the internal organs of the body. The phragmocone had a siphuncle whose purpose was to fill the chambers with gas or fluids to achieve neutral buoyancy in the water in the same manner as in ammonites and nautili. The order appeared in the Triassic, but flourished throughout the seas of the world during the Mesozoic Epoch, and died out in the Eocene.

FAMILY BELEMNITELLIDAE Pavlow, 1913

The shell consists of a pen-shaped calcitic rostrum or guard that covers the gas and fluid chambers of the phragmocone. The family closely follows the description of the order, characterized by a large phragmocone and a rapidly tapering rostrum. Members of the Family Belemnitellidae have worldwide distribution in rocks of the Cretaceous and Cenomanian through the Maastrichtian.

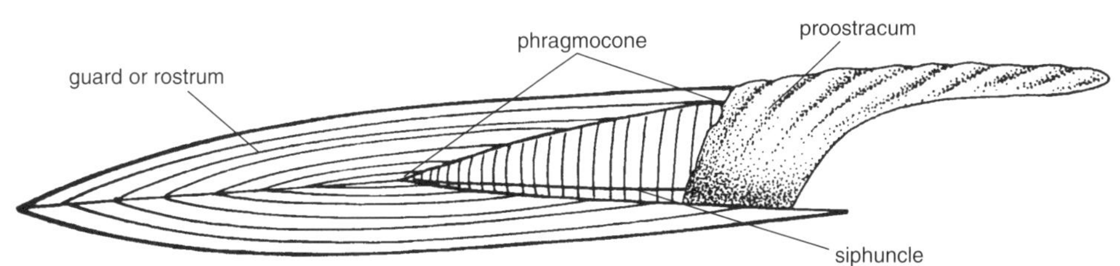

A Specialized Belemnite Cross Section with Unsectioned Proostracum

Genus *Belemnitella* d'Orbigny, 1840
belemnon = a dart + *telos* or *tela* = end

The genus follows the description of the order and the family very closely. However, there is a lengthwise slit on the ventral side of the rostrum and a flattened ridge on the dorsal side. The phragmocone is distinguished by a minute bulb or protoconch at the apex or tip. The genus occurs sparingly throughout the Upper Cretaceous of the Atlantic Coast, Gulf Coast, and the Western Interior Basin.

Belemnitella bulbosa Meek and Hayden, 1856

The rostrum is slender, subcylindrical, and slightly expanded at the anterior end. It has a moderately distinct dorsal ridge and a phragmocone. The phragmocone is as long or slightly longer than the rostrum, but is rarely completely preserved. Its final chambers are greater in diameter than the rostrum. The phragmocone tapers at an angle of 20° with an oval cross section. The septal walls curve slightly forward on the dorsal side. The species is found occasionally in the Upper Pierre Shale and more abundantly in the Fox Hills Formation.

Belemnitella bulbosa
BHI 4142
9.4 cm long phragmacone

ORDER NAUTILOIDEA

The Order Nautiloidea is distinguished by a curved to coiled external aragonitic shell containing a phragmocone and a body chamber. The phragmocone is segmented into many chambers by nearly straight septa and sutures. The nautiloids had a thicker shell than that of their relatives, the ammonites, which may have enabled them to withstand crushing at deep oceanic depths and pressures. The chambers are all connected as in Ammonoidea and Belemnitidea by a tube called a siphuncle that provided gas or fluids to the chambers. Nautiloids are and were relatives of the squid and octopi, and resided in a hard shell that provided protection for their soft parts as well as buoyancy for their body mass.

Extant *Nautilus* inhabit oceanic coastal areas to a depth of approximately 300 m. At night, they move up to a depth of approximately 100 m to feed, breed, and lay eggs. Cretaceous Age nautiloids appear to have inhabited water of 100 m or less during most of their lifespan. Differences between extant and Cretaceous Age nautiloids appear also in their egg-laying capacity. Present-day *Nautilus* lay about 10 eggs a year whereas their ancestral cousins laid smaller eggs, probably in greater quantity, estimated at between 10 and 50 eggs per year (personal communication, Landman 1994).

Today's *Nautilus* prefer cool or moderate water temperatures between 15°C and 22°C. Water temperatures of greater than 27°C can kill the nautilus (Saunders and Landman, 1987). Extant *Nautilus* also have a very low metabolic rate and may go for weeks without eating while remaining capable of very rapid swimming. These same characteristics probably held true for the nautiloids of the Cretaceous since today's *Nautilus* appears to be closely related to the *Eutrephoceras* of North America. This order has had worldwide distribution and has been found in rocks from the mid-Devonian Period through to the present. The continuing presence of this order allows the paleontologist to study a living fossil.

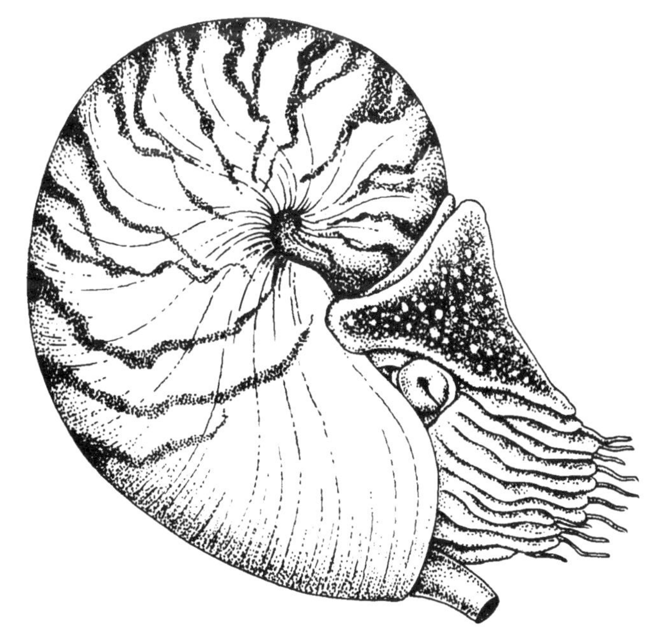

A Living (extant) Nautilus
Illustration by D.S. Norton

The Family Nautilidae is briefly described as having a generally smooth shell, which is involute to slightly evolute. The whorl section is slightly compressed. The siphuncle is located in the center (along the median line of symmetry), but may vary ventrally or dorsally within the chambers. The suture is straight to sinuous. Nautilidae are found from the Upper Triassic to the Recent and on every continent.

Genus *Eutrephoceras* Hyatt, 1894

eutrephas = nourishing + *ceras* = horn

The shell of *Eutrephoceras* is smooth and generally subglobose or nearly rounded. The whorl section is kidney shaped and broadly rounded on the flanks and venter. The aperture is slightly curved. The shell is involute and the suture pattern is slightly sinuous. The placement of the siphuncle is the primary characteristic for identifying species within the Pierre Seaway. This genus is present from the Upper Jurassic through the Miocene in North America and Colombia, South America.

Eutrephoceras alcesence Reeside, 1927

Eutrephoceras alcesence has a broadly rounded, stout shell, globose in juvenile stages to subglobose in the adult stage. The whorl section is nephritic (kidney shaped) and the placement of the siphuncle is central to dorsocentral. The surface of the shell has distinct longitudinal lines on the venter and coarse growth lines on the remainder of the shell. This species attained a large size, reaching a diameter of 24 cm. The ratio of height to width in the cross section ranges from 6 to 7. *Eutrephoceras alcesence* is present in the Eagle Sandstone of Montana, the Telegraph Creek Formation, and the Upper Cody Shale of the Big Horn Basin in Montana and Wyoming, the Steele Shale of east central Utah and New Mexico, and the basal Mesaverde Formation in the upper Rio Grande region of New Mexico. Another species, *Eutrephoceras thomi* Reeside, 1927, has been reported very sparingly from the Eagle Sandstone of Montana, but the authors believe this is probably synonymous with *Eutrephoceras alcesense,* and that the differences are only dimorphic.

Eutrephoceras dekayi (Morton, 1834)

Eutrephoceras dekayi has a broadly rounded, subglobose shell that is involute. The shell expands rapidly, more than doubling in size with each complete whorl. The height to width ratio of this species has been calculated at 3 to 4. The septa are reniform or kidneyshaped and the siphuncle is located one-third to one-fourth the distance from the dorsum or the inner wall. The shell surface of adult and medium-sized specimens is smooth with faint growth lines and slight longitudinal furrows or costae. The species is recorded from the Atlantic Coast, the Gulf Coast states, and the Pierre Shale and marine equivalents of the Western Interior from the *Scaphites hippocrepis* through the *Baculites compressus* Range Zones.

Side view
Eutrephoceras dekayi
BHI 4152
3.5 cm high

Venter view
BHI 4151
6.5 cm across

Eutrephoceras "elegans" var. *nebrascensis* (Meek and Hayden, 1862)

The description for *Eutrephoceras "elegans"* var. *nebrascensis* is a subglobose shell, broadly rounded on the flanks and venter. The umbilicus is closed in young and medium-sized specimens, and the shell is involute at larger sizes. The whorl increases rapidly in size with about the same dimensions in width and height. The height to width ratio is 3.5 to 3. The suture has a slight curve bending forward near the umbilicus, slightly backward on the sides, and slightly forward on the venter or periphery. The surface of the shell is smooth with distinct lines of growth and slight longitudinal costae. The placement of the siphuncle is about one-third the distance from the outside periphery. *Eutrephoceras "elegans"* var. *nebrascensis* has been noted from the *Didymoceras cheyennense* through *Baculites eliasi* Range Zones and appears to be prolific within the *Baculites compressus* and *Baculites cuneatus* Range Zones of the Western Interior of North America. The authors believe that the variety *nebrascensis* should be elevated to species status. The species *elegans* is an English Cenomanian stage *Nautilus* of a different subfamily and genus that does not occur in the Upper Cretaceous of North America.

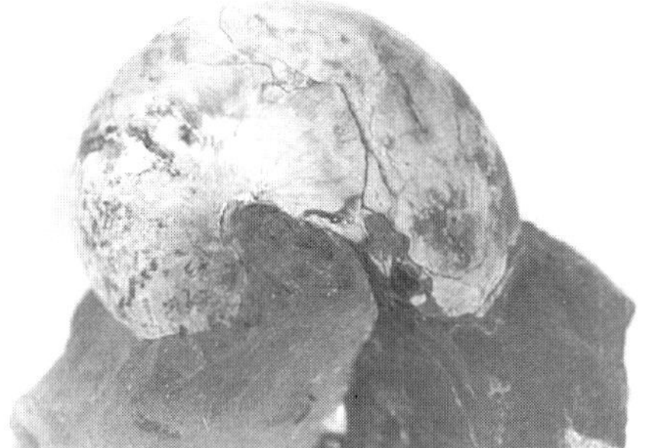
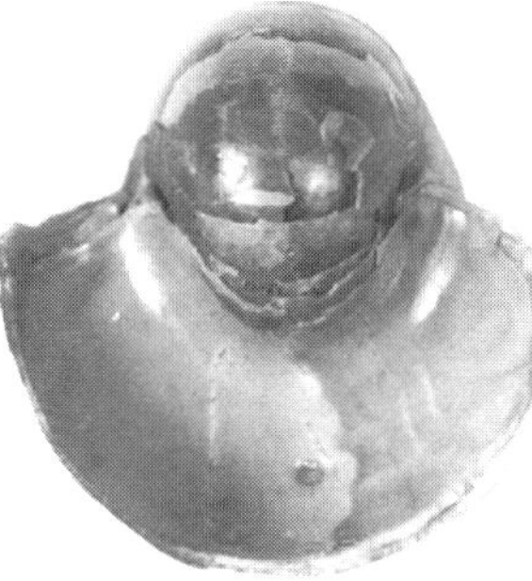

Left: Side view *Eutrephoceras "elegans"* var. *nebrascensis*
BHI 4154 6.5 cm high

Right: Venter view BHI 4153 4 cm across

Eutrephoceras dekayi var. *montanaensis* (Meek, 1876)

Eutrephoceras dekayi var. *montanaensis* has a subglobose, smooth shell. Whorl width does not generally increase as fast, allowing the height to be slightly greater than in the earlier forms. In this species, the height to width ratio is 4 to 3.5. Meek (1876) believed that this variety should be a separate species and the authors agree with this diagnosis. The placement of the siphuncle near the center of the chambers is the most distinguishing feature of this nautiloid. *Eutrephoceras dekayi* var. *montanaensis* is most abundant within the *Baculites baculus* Zone, but also appears slightly earlier and may range into the early Fox Hills Formation.

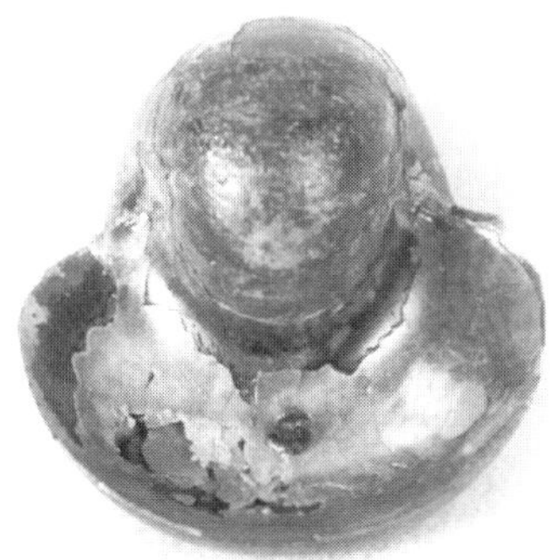
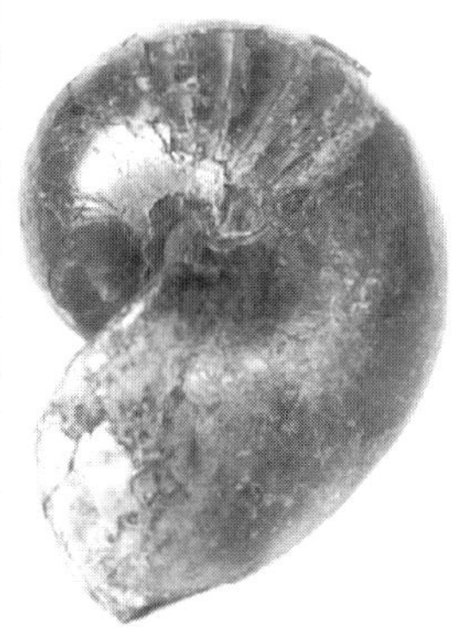

Eutrephoceras dekayi var.
montanaensis
BHI 4156
5.6 cm high

Eutrephoceras dekayi var.
montanaensis
BHI 4155
11 cm high

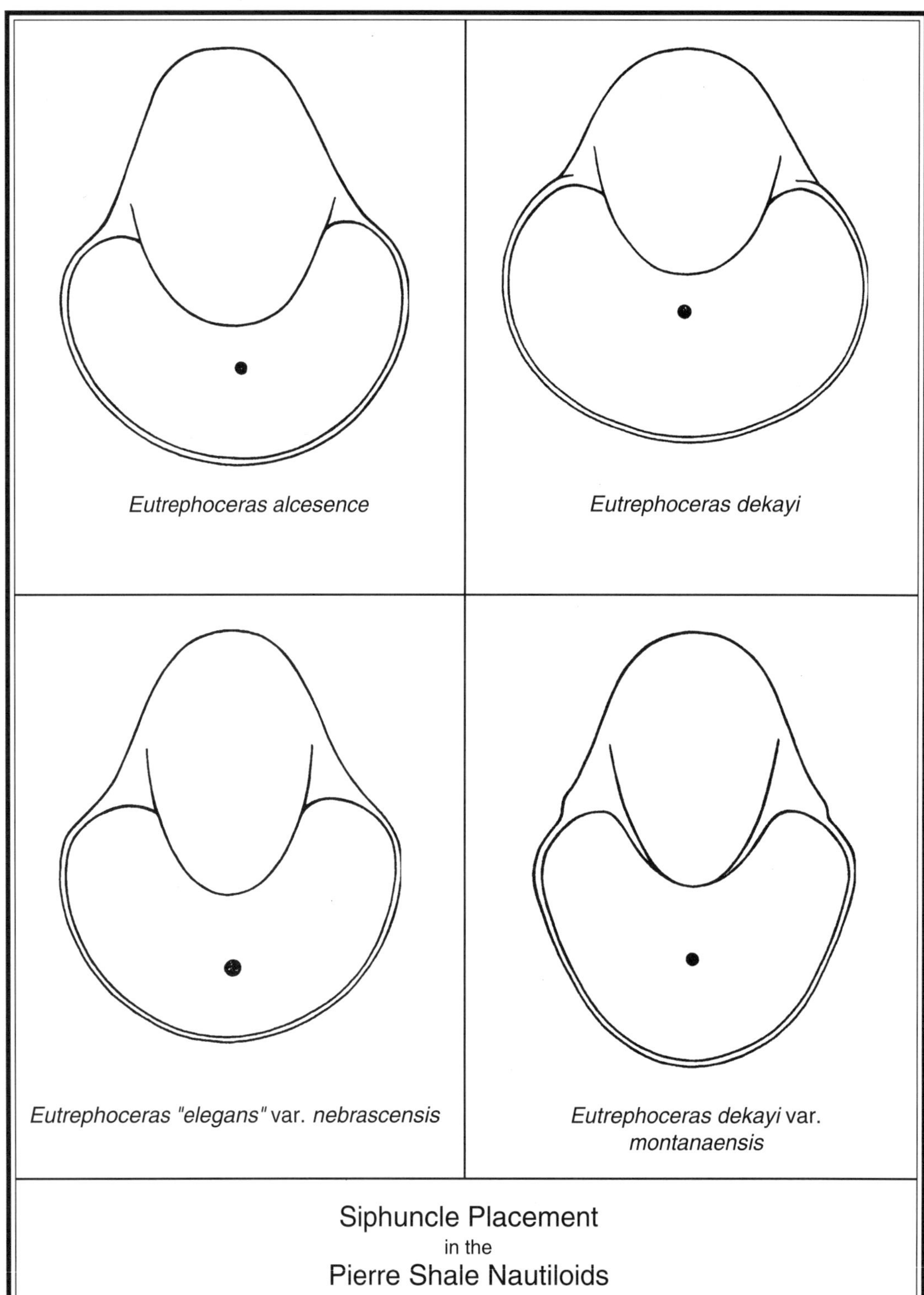

Illustrations by D.S. Norton

ORDER TEUTHIDEA

Order Teuthidea, the squidlike coleoids, is characterized by a reduced, inner shell that lacks a phragmocone. Young individuals of these recent and fossil coleoids have vestiges of phragmocone-like septa that lack a siphuncle. This proostracum forms the principal part of the commonly calcified horny exoskeleton or gladius of the coleoid that has a spoonlike or rounded shape at its anterior end. The preserved gladius consisted of a hard, inner shell on the dorsum or top of the squid that served as a protective plate for the soft internal organs. The Teuthidea have wings or fins that vary in size and shape and are found protruding from the dorsal parts of the proostracum. Fossil and extant teuthids have four arm pairs and a pair of tentacles, although the fossil teuthids and some recent forms sometimes lack hooks and have suckers, or suckers and feelers, on their tentacles. Today's squid, which inhabit coastal waters, are the direct descendants of earlier coleoids. Order Teuthidea has been found from the Jurassic Age to the present.

FAMILY KELAENIDAE Naef, 1921
(see Jeletzky, 1966)

The Family Kelaenidae has a simple characteristic shape described thus: "the end of the pen is narrow and forms a guard and the posterior end is enlarged to form a gladius that spreads out laterally into the conus" (Jeletzky, 1966). The Family Kelaenidae occurs worldwide from the Middle Jurassic through the Upper Cretaceous.

Genus *Tusoteuthis* Logan, 1898
tuso = ? + *teuthis* = squid

According to Karl Waage (personal communication, 1994), the genus follows the description of the family, but differs greatly from the original Kelaeno (Jurassic). *Tusoteuthis* is found within the *Baculites obtusus* through the *Baculites perplexus* Range Zones of the Western Interior. Three specimens of a new species, and possibly a new genus, that have been reported from the Wasta, South Dakota area, originate from the *Baculites compressus* through the *Baculites reesidei* Zones of the Pierre Shale.

Tusoteuthis sp.
BHI 4138
21 cm long

Tusoteuthis longa Logan, 1898
(see also Nicholls and Isaak, 1987)

The gladius of *Tusoteuthis longa* is shaped like a spear point directed toward the posterior end of the animal. The shell is composed of fibrous radiating calcite in the conus, rhachis, and gladius. The rhachis converges in the shape of a pen with the point extending about halfway onto the gladius. The conus is represented by a V-shaped anterior end and a broad rounded swelling over the posterior end of the gladius. The rhachis served as a guard over the mantle to provide support. The gladius acted as protection and support for the soft organs in the squid. The posterior end of the gladius also formed a base for the steering fins. The species has been found in the Sharon Springs Member, the Gammon Ferruginous Member of the Pierre Shale in Wyoming, Nebraska, and South Dakota, the Pembina Member of the Pierre Shale in southern Manitoba, and the Smokey Hill Chalk of Kansas.

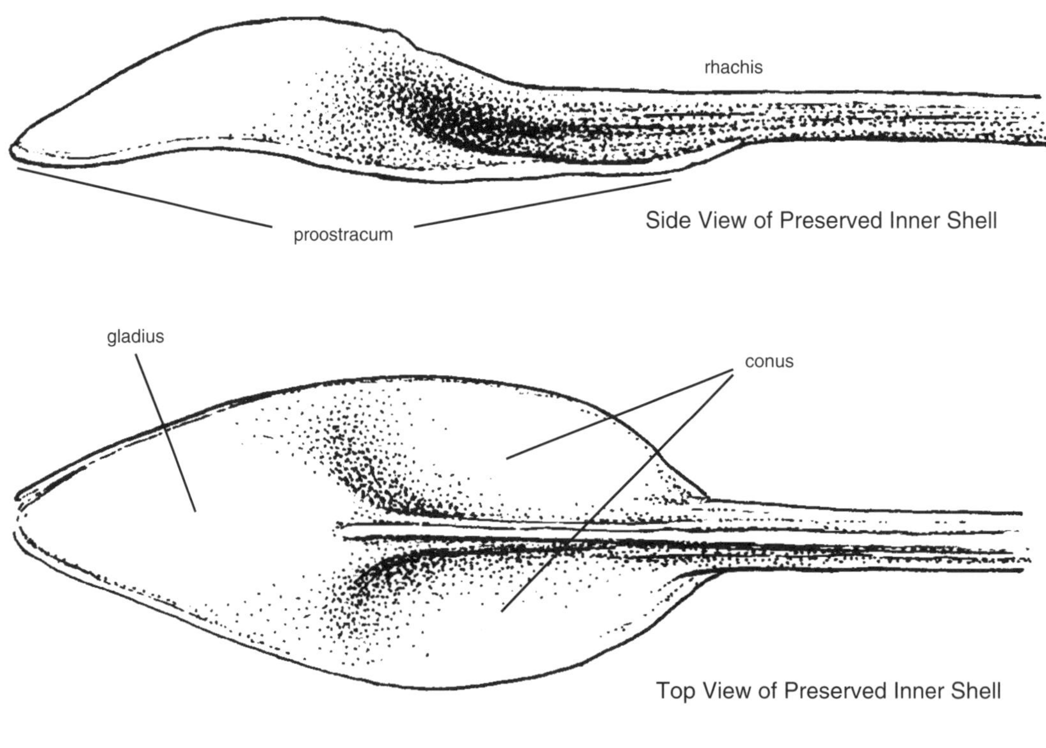

Tusoteuthis sp.
Drawn from BHI 4138
Illustrations by D.S. Norton

Trachyteuthids have a broad, sturdy, cuttlefish-like internal gladius that is strongly calcified and exhibits a knobby, dorsal texture. The gladius side and middle plates are curved and slightly offset, forming asymptotes; the coni (veins) are about one-half of the gladius length. The family has a worldwide distribution in the Jurassic and Cretaceous.

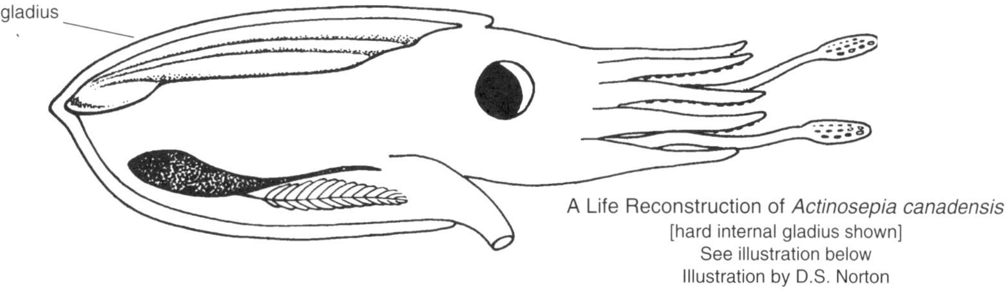

A Life Reconstruction of *Actinosepia canadensis*
[hard internal gladius shown]
See illustration below
Illustration by D.S. Norton

Genus *Actinosepia* Whiteaves, 1897
aktinos = a ray + *sepia* = cuttlefish

Actinosepia canadensis Whiteaves, 1897

Actinosepia has a broad, oval, arched gladius that is concave on the venter. The gladius expands forward like a fan from a rounded conus and reaches its maximum width at about one-third the length from the anterior end where it then remains nearly straight. Five longitudinal ribs radiate from the apex with the middle rib being the longest and the two outside ribs being the shortest. The ribs project beyond the body of the gladius, giving the appearance of a duck's foot. The ventral side of the gladius is smooth and the dorsal view is ornamented with fine to coarse tubercles. *Actinosepia canadensis* is found in the Upper Campanian and Maastrichtian of Canada and in the United States from Alaska to Texas. The species has a length about $2^{1}/_{2}$ times the maximum width, or 18 to 30 cm in adult specimens.

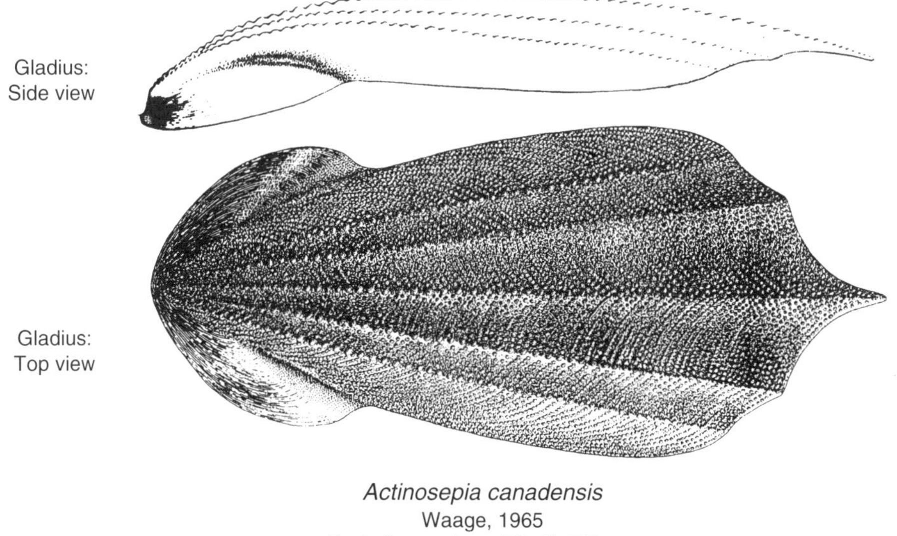

Actinosepia canadensis
Waage, 1965
Illustration courtesy of Dr. Karl Waage

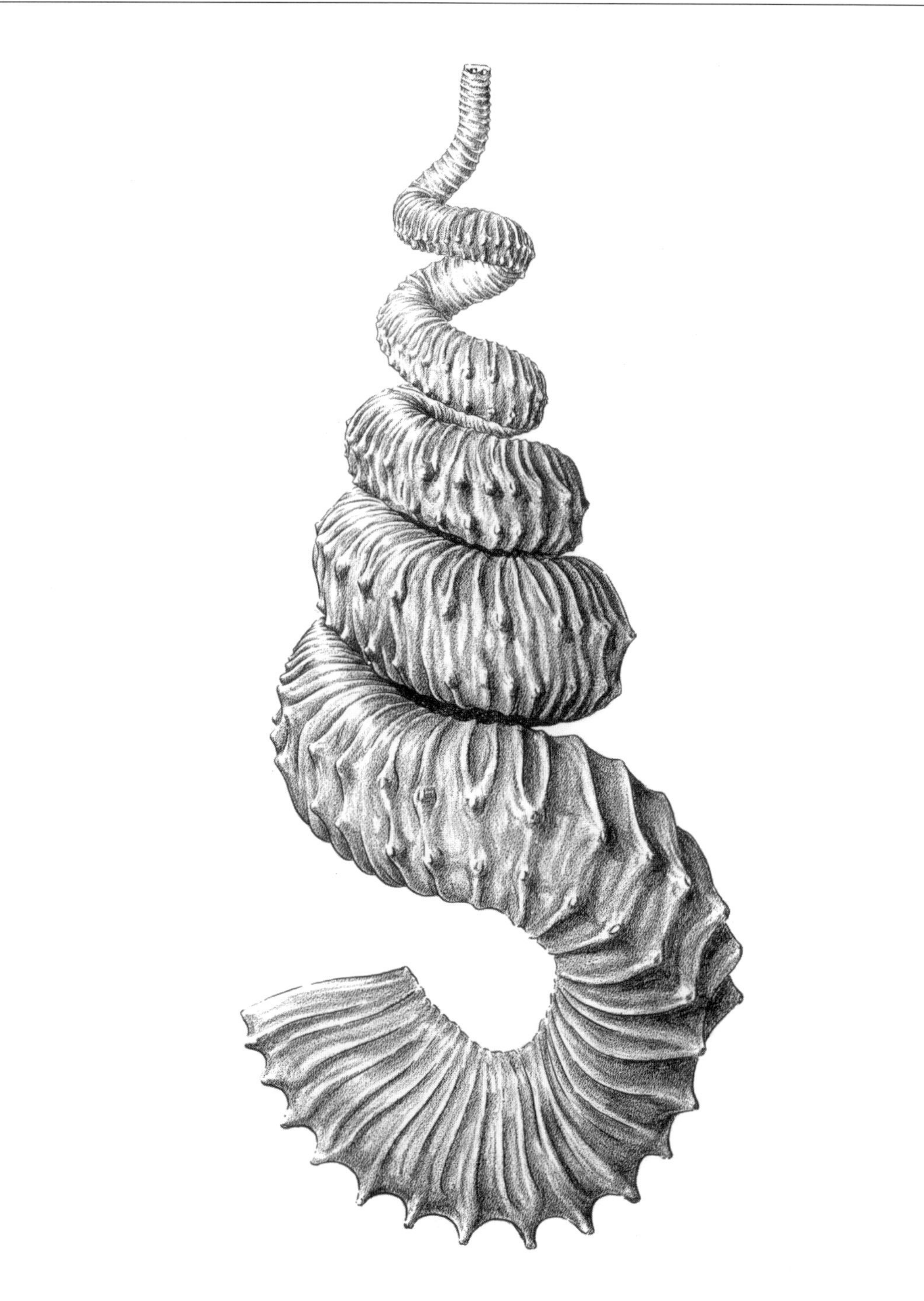

Didymoceras stevensoni
Drawing by John R. Stacy

PHYLUM ANNELIDA
- *Serpula cretacea*
- *Serpula markmani*
- *Serpula* sp.
- *Omasaria* sp.
- *Diploconcha* sp.

PHYLUM ARTHROPODA
- Class Crustacea
- Order Decapoda
  - *Callianassa cheyennensis*
  - *Dakoticancer overanus*
  - *Dioratiopus heartailensis*
  - *Dioratiopus dawsonensis*
  - *Dromiopsis kimberlyae*
  - *Ekalakia lamberti*
  - *Eomunidopsis cobbani*
  - *Glaesnerella* sp.
  - *Homolopsis dawsonensis*
  - *Homolopsis punctata*
  - *Homolopsis mendryki*
  - *Hoploparia bearpawensis*
  - *Hoploparia dawsonensis*
  - *Hoploparia mickelsoni*
  - *Linuparus pustulosus*
  - *Necrocarcinus pierrensis*
  - *Necrocarcinus davisi*
  - *Necrocarcinus labeschei*
  - *Notopocorystes (Eucorystes) eichhorni*
  - *Palaeonephrops browni*
  - *Plagiophthalmus* sp.
  - *Protocallianassa russelli*
  - *Raninella oaheensis*
  - *Rugafarius fredrichi*
  - *Sodakus tatankayotankaensis*
  - *Tetracarcinus subquadratus*
  - *Xanthosia elegans*
  - *Xanthosia elegans occidentalis*
  - *Zygastrocarcinus griesi*
  - *Zygastrocarcinus cardsmithi*
  - *Zygastrocarcinus mendryki*

PHYLUM BRACHIOPODA
- Class Inarticulata
- Order Lingulida
  - *Lingula subspatulata*

PHYLUM BRYOZOA
- Pyriporoid bryozoan
- Membraniporoid bryozoan

PHYLUM COELENTERATA
- Class Anthozoa
- Order Scleractinia
  - cf. *Websteria* sp.
  - *Micrabacia americana*
  - *Micrabacia ?* sp.
  - *Trochocyathus* sp.

PHYLUM ECHINODERMATA
- Class Echinoidea
- Order Cassiduloida
  - *Hardouinia taylori*
- Order Spatangoida
  - *Hemiaster humphreysanus*
  - *Hemiaster beecheri*
- Order Cidaroida
  - *Eurysalenia minima*

PHYLUM MOLLUSCA
- Class Cephalopoda
- Order Ammonoidea
  - *Anaklinoceras reflexum*
  - *Anaklinoceras gordiale*
  - *Anaklinoceras* sp.
  - *Axonoceras compressum*
  - *Baculites aquilaensis*
  - *Baculites haresi*
  - *Baculites* sp. smooth (Cobban)
  - *Baculites* sp. weak flank ribs (Cobban)
  - *Baculites obtusus*
  - *Baculites mclearni*
  - *Baculites asperiformis*
  - *Baculites* smooth species (Cobban)
  - *Baculites gilberti*
  - *Baculites perplexus*
  - *Baculites gregoryensis*
  - *Baculites reduncus*
  - *Baculites scotti*
  - *Baculites* sp. (new species)
  - *Baculites pseudovatus*
  - *Baculites crickmayi*
  - *Baculites rugosus*
  - *Baculites corrugatus*
  - *Baculites compressus*
  - *Baculites compressus* var. *robinsoni*
  - *Baculites* sp.
  - *Baculites cuneatus*
  - *Baculites reesidei*
  - *Baculites undatus*
  - *Baculites jenseni*
  - *Baculites eliasi*
  - *Baculites baculus*
  - *Baculites grandis*
  - *Baculites clinolobatus*
  - *Cirroceras conradi*
  - *Coahuilites sheltoni*

Didymoceras cochleatum
Didymoceras mortoni
Didymoceras tortum
Didymoceras binodosum
Didymoceras cf. *archiacianum*
Didymoceras nebrascense
Didymoceras stevensoni
Didymoceras cheyennense
Didymoceras sp.
Glyptoxoceras rubeyi
Exiteloceras jenneyi
Haresiceras natronense
Haresiceras placentiforme
Hoplitoplacenticeras cf. *H. coesfeldiense var. schlüteri*
Hoplitoplacenticeras marroti
Hoploscaphites gilli
Hoploscaphites landesi
Hoploscaphites nicolletii var. saltgrassensis
Hoploscaphites melloi
Hoploscaphites birkelundi
Jeletzkytes nodosus
Jeletzkytes brevis
Jeletzkytes quadrangularis
Jeletzkytes cf. *nodosus*
Jeletzkytes furnivali
Jeletzkytes plenus
Jeletzkytes crassus
Jeletzkytes criptonodosus
Jeletzkytes dorfi
Menabites (Delawarella) danei
Menabites (Delawarella) vanuxemi
Menuites oralensis
Menuites portlocki complexus
Nostoceras aff. *N. humile*
Nostoceras aff. *N. colubriformis*
Nostoceras monotuberculatum
Oxybeloceras sp.
Oxybeloceras crassum
Oxybeloceras meekanum
Pachydiscus cf. *oldhami*
Pachydiscus arkansanus
Pachydiscus catarinae
Pachydiscus cf. *hornbyense*
Parapuzosia bradyi
Parasolenoceras pulcher
Placenticeras sancarlosense
Placenticeras guadalupae
Placenticeras planum
Placenticeras pingue
Placenticeras intercalare
Placenticeras meeki
Placenticeras costatum
Ponteixites gracilis
Ponteixites robustus

Pseudobaculites natosoni
Rhaeboceras subglobosum
Rhaeboceras coloradoense
Rhaeboceras mullananum
Rhaeboceras burkholderi
Rhaeboceras halli
Rhaeboceras albertense
Scaphites hippocrepis I
Scaphites hippocrepis II
Scaphites hippocrepis III
Solenoceras crassum
Solenoceras texanum
Solenoceras mortoni
Solenoceras n. spp.
Solenoceras reesidei
Sphenodiscus pleurisepta
Submortoniceras tequesquitense
Trachyscaphites pulcherrimus
Trachyscaphites redbirdensis
Trachyscaphites spiniger porchi
Trachyscaphites praespiniger
Order Belemnitidea
    Belemnitella bulbosa
Order Nautiloidea
    Eutrephoceras alcesense
    Eutrephoceras thomi
    Eutrephoceras dekayi
    Eutrephoceras "elegans" var. nebrascensis
    Eutrephoceras dekayi var. montanaensis
Order Teuthidea
    Actinosepia canadensis
    Tusoteuthis longa
Class Gastropoda
    Acirsa (Hemiacirsa) n. sp.
    Acmaea occidentalis
    Aceton cf. *throckmortoni*
    Akera ? sp.
    Amuletum minor
    Amauropsis ? sp.
    Anchura nebrascensis?
    Anchura haydeni
    Anchura sublevis
    Anchura ? *parva*
    Anisomyon borealis
    Anisomyon centrale
    Anisomyon patelliformis
    Anisomyon subovatus
    Anisomyon alveolus
    Anisomyon sexsulcatus
    Anomia sp.
    Anomalofusus ? sp.
    Astandes densatus
    Aporrhais biangulata

Atira ? *nebrascensis*
Bellifusus ? n. sp.
Belliscala ? n. sp.
Bullopsis aff. *B. cretacea*
Bullopsis n. sp.
Capulus spangleri
Cerithioderma n. sp.
Cerithiopsis (Cerithiella) n. sp.
Closteriscus tenuilineatus
Cryptorhytis cheyennensis
Cryptorhytis flexicostata
Cylichna cf. *secalina*
Cylichna cf. *incisa*
Cylichna sp.
Drepanochilus evansi
Drepanochilus scotti
Drepanochilus nebrascensis
Drepanochilus obesus
Drepanochilus sp.
Ellipsoscapha occidentalis
Ellipsoscapha subcylindrica
Ellipsoscapha aff. *E. occidentalis*
Ellipsoscapha aff *E. subcylindrica*
Ellipsoscapha n. sp.
Eoacteon n. sp.
Euspira obliquata
Fasciolaria ? *gracilenta*
Graphidula culbertsoni
Graphidula cf. *alleni*
Graphidula cf. *obscura*
Gyrodes cf. *spillmani*
Gyrodes subcarinatus
Lomirosa n. sp.
Lunatia sp.
Lunatia subcrassa
Margarita nebrascensis
Margaritella flexistriata
Medionapus ? sp.
Mesorhytis gracilenta
Natica (Lunatia) concinna
Nonacteonina sp.
Nonacteonina attenuata
Oligoptycha concinna
Oligoptycha sp.
Paladmete n. sp.
Polinices concinna
Polinices rectilabrum
Polinices sp.
Potamides n. sp.
Promathilda (Clathrobaculus) n. sp.
Pseudomaura paludinaeformis
Pseudobuccinum nebrascense
Pyrifusus ? sp.
Pyropsis ? sp.

Pyropsis ? n. sp.
Pyrifusus (Neptunella) newberryi
Pyrifusus (Neptunella) intertextus
Remera cf. *stephensoni*
Rhombopsis ? sp.
Rhombopsis newberryi
Rhombopsis subturritus
Rhombopsis ? *intertextus*
Scobinodolus sp.
Serrifusus dakotensis
Spironema ? sp.
Tornatellaea cretacea
Trachytriton vinculum
Turris (Surcula) minor
Turritella ? sp.
Vanikoropsis nebrascensis
Vanikoropsis haydeni
Vanikoropsis tuomeyana
Volutoderma ? *clayworthyi*
Xenophora sp.
Class Pelecypoda
Agerostrea mesenterica
Anatina doddsi
Anatina sp.
Anomia flemingi
Anomia raetiformis
Anomia argentaria
Anomia subtrigonalis
Anomia cf. *argentaria*
Anomia tellinoides
Anomia sp.
Anomia n. sp.
Aphrodina ? sp.
Astarte gregaria
Astarte sp.
Clisocolus moreauensis
Corbula crassimarginata
Crassatella subquadrata
Crassatella sp.
Crassostrea glabra
Crenella cf. *elegantula*
Crenella aff. *C. microstriata*
Crenella sp.
Cuspidaria grovensis
Cuspidaria variabilis
Cuspidaria ventricosa
Cuspidaria moreauensis
Cuspidaria sp.
Cyclorisma ? sp.
Cymbophora warrenana
Cymbophora canonensis
Cymbophora holmesi
Cymbophora gracilis
Cymbophora sp.

Cymbophora n. sp.
Dosiniopsis deweyi
Ethmocardium welleri
Exogyra costata
Exogyra sp.
Gervillia sp.
Goniomya americana
Goniomya sp.
Goniochasma stimpsoni
Idonearca shumardi
Inoceramus pertenuis
Inoceramus oblongus
Inoceramus sublaevis
Inoceramus agdjakendensis
Inoceramus subcompressus
Inoceramus aff. *I. proximus*
Inoceramus azerbaidjanensis
Inoceramus convexus
Inoceramus barabini
Inoceramus aff. *I. turgidus*
Inoceramus sublaevis
Inoceramus sagensis
Inoceramus tausiensis
Inoceramus cf. *I. shikotanensis*
Inoceramus tenuilineatus
Inoceramus proximus
Inoceramus canadensis
Inoceramus vanuxemi
Inoceramus aff. *I. pertenuis*
Inoceramus mcshaniensis
Inoceramus furnivali
Inoceramus palliseri
Inoceramus subcircularis
Inoceramus cf. *I. proximus*
Inoceramus mclearni
Inoceramus cf. *I. nebrascensis*
Inoceramus (Endocostea) sulcatus
Inoceramus cf. *I. balticus*
Inoceramus typicus
Inoceramus incurvus
Inoceramus cf. *balchi*
Inoceramus fibrosus
Leda pittensis
Leda hindi
Legumen ellipticum
Legumen sp.
Lima pelagica
Limatula aff. *L. acutilineata*
Limopsis sp.
Limopsis parvula
Lucina occidentalis
Lucina subundata
Modiolus aff. *M. wenonah*
Modiolus uddeni

Modiolus meeki
Modiolus cf. *wrighti*
Nemodon sulcatinus
Nemodon adkinsi
Nemodon eufaulensis
Nemodon sp.
Nucula cancellata
Nucula planimarginata
Nucula nacatochana
Nucula subplana
Nucula cf. *subplana*
Nucula (Pectinucula) n. sp.
Nuculana bisulcata
Nuculana evansi
Nuculana corbetensis
Nuculana corsicana
Nuculana sp.
Nymphalucina subundata
Nymphalucina occidentalis
Nymphalucina sp.
Nymphalucina n. sp.
Opertochasma cuneatum
Ostrea russelli
Ostrea cf. *O. russelli*
Ostrea sp.
Ostrea plumosa
Ostrea cf. *O. falcata*
Ostrea inornata
Oxytoma haydeni
Oxytoma nebrascana
Pachymya ? *aurandi*
Panope berthoudi
Panope sp.
Pecten (Chlamys) nebrascensis
Pecten (Syncyclonema ?*) simplicius*
Pecten (Amusium ?*)* cf. *danei*
Pecten (Camptonectes) sp.
Periploma ? sp.
Perrisonota protexta
Phelopteria linguaeformis
Phelopteria sublevis
Pholadomya hodgei
Pholadomya sp.
Pinna lakesii
Pinna sp.
Protocardia rara
Protocardia subquadrata
Protocardia sp.
Pseudoptera ? sp.
Pteria petrosa
Pteria cf. *parkensis*
Pteria linguaeformis
Pteria (Oxytoma) nebrascana
Pteria sp.

Solemya n. sp.
Spyridoceramus fibrosus
Syncyclonema halli
Syncyclonema n. sp.
Tancredia sp.
Tellina munda
Tellinimera scitula
Tenea sp.
cf. *Tenuiptera* sp.
Thetiopsis circularis
Thracia n. sp.
Thyasira rostrata rostrata
Thyasira rostrata cracens
Thyasira quadrula quadrula
Thyasira quadrula arrecta
Thyasira triangulata
Thyasira beauchampi beauchampi
Thyasira beauchampi rex
Thyasira advena advena
Thyasira becca becca
Thyasira becca cobbani
Thyasira cantha
Thyasira n. sp.
Veniella aff. *V. conradi*
Veniella sp.
Veniella humilis
Yoldia scitula
Yoldia evansi
Class Scaphopoda
  Cadulus obnutus
  Dentalium gracile
  Dentalium pauperculum
  Dentalium sp.

PHYLUM CHORDATA
Class Chondrichthyes
  Order Lamniformes
    Cretolamna appendiculata
    Odontaspis sp.
    Squalicorax cf. *S. kaupi*
  Order Hybodontiformes
    Ptychodus sp.
  Order Chimaeriformes
    Ichthypriapis gladius
Class Osteichthyes
  Order Pachycormiformes
    Protosphyraena gladius
    Protosphyraena gigas
  Order Ichthyodectiformes
    Gillicus arcuatus
    Ichthyodectes ctenodon
    Xiphactinus audax
    ichthyodectid undet.
    Saurocephalus lanciformis
    Saurodon leanus

"Prosaurodon" pygmaeus
Bananogmius evolutus
Order Elopiformes
  Palaeoclupea dakotaensis
  Eloposis sp.
  Apsopelix minimus
  Apsopelix berycinus
  Pachyrhizodus caninus
  Pachyrhizodus minimus
Order Aulopiformes
  Cimolichthys nepaholica
  Enchodus petrosus
  Enchodus gladiolus
  Stratodus apicalis

Class Reptilia
  Order Chelonia
    Toxochelys latiremis
    Archelon ischyros
    Archelon marshi
    Protostega sp.
  Order Plesiosauria
    Dolichorhynchops osborni
    Polycotylus latipinnus
    Elasmosaurus platyurus
    Styxosaurus browni
    Hydralmosaurus serpentinus
    Alzadasaurus pembertoni
  Order Squamata
    Platecarpus cf. *P. somenensis*
    Platecarpus ictericus
    Platecarpus sp.
    Plioplatecarpus primaevus
    Plioplatecarpus sp.
    Prognathodon crassartus
    Prognathodon overtoni
    Mosasaurus missouriensis
    Mosasaurus conodon
    Clidastes propython
    Globidens dakotensis
    Tylosaurus proriger
    Hainosaurus sp.
  Order Pterosauria
    Pteranodon sp.
  Order Saurischia
    ornithomimid gen. undet.
  Order Ornithischia
    hadrosaurid gen. undet.
    ceratopsian gen. undet.
Class Aves
  Order Hesperornithiformes
    Hesperornis regalis
    hesperornithiform undet.
  Order Ichthyornithiformes
    Ichthyornis sp.

aberrant. wandering; straying from the normal or typical form.

acute (periphery). with sides of shell meeting at sharp angle without shoulders.

adapical. toward apex or ammonitella of the shell; backward direction.

adoral. toward mouth of ammonoid or aperture of shell; forward direction.

aff. an abreviation for affinity meaning an inherent similarity.

ammonitella. the initial hatchling of an ammonite that consists of a coiled protoconch, one or possibly more septa, and a body chamber. The first septum is composed entirely of prismatic rather than nacrous aragonite.

aperture. open end of body chamber of shell, where the head and tentacles of the animal extended from the shell.

approximated (ribs). crowded toward aperture, usually associated with maturity of growth.

approximated (sutures). crowded toward body chamber, usually indicating maturity.

aptychus. pair of plates serving as jaws and possibly for closure of aperture in some ammonoids; single plate (anaptychus) or pair of plates (aptychus).

aragonite. a crystalline form of calcium carbonite, one of the constituents of mollusc shells.

arcuate (ribs). to arch, bend, or curve like a bow.

asymptote. a line that continually approaches nearer to a curve but never reaches it.

auxiliary (lobe or saddle). lateral lobe or saddle of suture springing from umbilical lobe or saddle between second lateral and umbilical seam.

benthic. the bottom or depths of the ocean.

bifurcate (rib). dividing into two branches toward venter.

body chamber. large undivided space in shell extending from the phragmocone to the aperture that was inhabited by the living animal (this space is also called the living chamber).

bulla (bullae, pl.; bullate, adj.). tubercle expanded radially.

bundled (ribs). united in bunches or sheaves at or near umbilical shoulder, usually at a tubercle.

camera (camerae, pl.). compartment between two adjacent septa comprising one of the spaces into which the entire shell between protoconch and body chamber is divided by the septa.

cf. an abbreviation for the Latin word meaning to confer or compare with. In paleontology, to compare similarities between this species and another species.

chamber. *See* body chamber, camera.

clavus (clavi, n., clavate, adj.) tubercle elongated in direction of coiling (longitudinally).

compressed (whorl section). higher than wide.

concave (rib). bowed away from aperture.

concave (side or venter). having a surface that curves in like the interior surface of a sphere or lens.

conch. a one-piece shell in mollusks, generally composed of aragonite and conchiolin.

constricted. drawn together; smaller or narrower in one place.

conus. a cone or cone-shaped shell or organ.

convex (rib). bowed toward the aperture.

convex (side or venter). having a surface that curves outward like the exterior surface of a sphere or arc.

corrugations. series of parallel folds, ridges, wrinkles, or furrows.

costae. ribs or parts resembling ribs, ridges, or shells.

cross section. section made by plane crossing through at a right angle to the axis of an object (i.e., such as the end view of a baculite).

dense (ribs). closely spaced; depressed (whorl section); wider than high.

dextral. right-handed coiling whorls descending to the aperture in clockwise spirals (Nostoceratidae).

Dibranchiata. a group of cephalopods, including the squid, octopi, and cuttlefish, that have two gills, an ink sac, and many armlike appendages.

dimorphism. the existence of two body shapes in the same species, usually as a result of gender.

distant (ribs). widely spaced.

dorsal lobe. median primary lobe of suture on the dorsum (internal in normally coiled conchs).

dorsum. dorsal side of conch (opposite ventral or siphuncle wall), generally grading into dorsolateral areas; in slightly involute shells equivalent to impressed area, but in deeply involute shells refers only to portion of conch adjacent to venter of preceding whorl.

endemic. prevalent in or restricted to a particular group, region, or locality.

endoskeleton. having an internal supporting skeleton or shell (i.e., vertebrates, squid).

exoskeleton. having an external protecting or supporting structure (i.e., insects, crustaceans).

extant. still in existence.

extinct. no longer existing.

evolute. with whorls overlapping little or not at all, and therefore having a wide umbilicus. (As commonly used, evolute and involute are relative terms, since a shell form called evolute in one family may be classed as involute in another.)

falcate (rib). sickle-shaped.

falcoid (rib). approaching sickle-shaped.

fasciculate (ribbing). with ribs bunched or bundled to form sheaves.

flank. lateral wall of whorl between umbilical seam and venter, also called **whorl side**.

formation. A body of rocks of intermediate rank in the hierarchy of lithostratigraphic units; it is defined and identified by its lithologic composition and stratigraphic position. A formation is the fundamental unit in formal lithostratigraphic classification.

flexuous (rib). bending or bent, changing direction in a curve.

gen. an abbreviation for genus.

gladius (pl. gladii). the horny endoskeleton or pen of a two-gilled cuttlefish.

globose. rounded, almost spherical.

helical or helicoidal. coiled in regular three-dimensional spiral form with constant spiral angle, as in most gastropods and nostoceratids.

heteromorph. ammonoid shell of any form except planispiral with whorls in contact.

hook. the curved or bent body chamber of the ammonite reserved for aberrant forms of ammonite (Scaphitidae and Nostoceratidae).

intercalatory (rib). secondary rib not attached to primary rib, at least on one side of whorl.

involute. with whorls overlapping considerably and hence with narrow umbilicus (*see* evolute).

keel. continuous distinct longitudinal ridge on venter (also referred to occasionally as venter).

lappet. simple or necked (spatulate) projection of peristome on whorl sides or venter (called ventral lappet when located on venter; called dorsal lappet when located on dorsum).

last septum. septum separating body chamber from adjoining camera at any stage of growth; adoral septum.

lenticular. shaped like a lentil, or a double convex lens.

lobe. element of suture directed backward toward the phragmocone.

mantle. membrous flap or folds of the body wall of a mollusk, containing the glands that secrete a shell-forming fluid.

member. the formal lithostratigraphic unit next in rank below a formation; it is always a part of a formation.

n. sp. an abbreviation for new species, used when species is first described within the present publication or when species description has not yet appeared in any publication.

nacreous. having an iridescent luster like mother-of-pearl.

node. large blunt or rounded tubercle.

ornament. features of shell exterior such as ribs, tubercles, bullae, clavi, spines, and striations.

pelagic. the ocean surface or open sea, distinguished from coastal waters.

period. the interval of geologic time during which the rocks of the corresponding system were formed.

periphery. the outer wall or venter of the shell.

peristome. the area or parts surrounding the mouth or opening.

phragmocone. divided or segmented (sutured area) part of shell.

planispiral. coiling all in one plane or horizon.

primary rib. main stem or simple inward part of a branched rib.

projected (rib). swung forward at or near venter.

proostracum. a bladelike or spoon-shaped plate located on the dorsum.

prorsiradiate (rib). with general forward inclination from umbilical side toward venter.

protoconch. first chamber or initial chamber of shell.

radula (radulae, pl.). a rasplike organ with toothlike processes in the mouth of some mollusks used for tearing food.

range zone. the body of strata representing the known stratigraphic and geographic range of occurrences of any selected element or elements of the assemblage of fossils present in a stratigraphic sequence.

rectiradiate (rib). in straight radial position, bending neither forward nor backward, like the spokes of a wheel.

retracted. the portion of a conch that no longer clasps around earlier growth stages, as in a scaphite or a baculite ammonitelle.

rhachis. the central portion of the radula or the structure supporting the radula.

rib. radially directed ridge on shell; sometimes called costa.

rostrum—guard. pointed projection of peristome on venter; may continue spiral line of coiling or diverge from it, or solid, pen-shaped structure forming the apex or point of the shell.

rursiradiate (rib). inclined backward proceeding from umbilical area toward venter saddle. (Element of suture directed forward.)

saddle. element of suture directed forward toward the body chamber.

sandstone. a sedimentary rock consisting of individual grains of sand-size particles 0.06 to 2 mm in diameter, either set in a fine-grained matrix (silt or clay) or bounded by chemical cement.

secondary rib. outer part of branched rib or a short rib between two longer primary ribs.

septum. transverse partition dividing shell into camerae, attached to inside of shell wall along suture line.

shaft. a long straight to slightly curved section of the whorl.

shale. a fine-grained laminated or fissile sedimentary rock made up of silt- or clay-size particles; generally consists of about one-third quartz, one-third clay materials, and one-third miscellaneous minerals, including carbonates, iron oxides, feldspars, and organic matter.

shell. complete hard parts of ammonoid, including protoconch and conch (but excluding aptychus and beaks or jaw structures, which generally are separated from the conch if preserved at all).

shoulder. umbilical or ventrolateral blunt angle of whorl.

simple (rib). unbranched.

simple (suture). not appreciably subdivided.

sinistral. left-handed coiling whorls descending to aperture in counterclockwise spirals (Nostoceratidae).

sinuous (suture). bent, curved, winding, bending in and out.

siphuncle. narrow longitudinal tube passing through camerae and septa from protoconch to base of body chamber.

sp. an abbreviation for species.

spine. a pointed projection extending from the shell.

strong (rib). very prominent, usually somewhat broad, separated from each other by a depression.

subglobose. not quite rounded.

suture. line of junction of septum with walls of shell, visible only when shell is removed; sometimes termed septal suture, suture line.

Tetrabranchiata. group of cephalopods comprising the nautiloids and ammonites; these cephalopods have four gills and an external shell.

trifurcate (rib). dividing into three branches.

trigonal. having three prominent, longitudinal angles or corners, triangular.

tubercle. projection or small, pointed swelling on shell surface, or an internal mold commonly representing the base of a spine.

unconformity. a surface of erosion between rock bodies, representing a significant loss or gap in the stratigraphic succession. Unconformities result from the exposure of rocks below the unconformity, and erosion, resulting in the loss of some parts or records of the older rocks.

umbilical area. inner part of whorl on each side, separating umbilical shoulder from umbilical seam; called **umbilical wall** if it rises vertically from the spiral plane, and **umbilical slope** if it rises gently.

umbilical margin. area of central coiling above and around the protoconch in involute and evolute ammonoids.

umbilical seam. contact in umbilicus of two adjoining whorls.

umbilical shoulder. generally blunt angle between flank and umbilical area, also called umbilical edge or angle.

umbilical width. diameter of umbilicus measured either between umbilical shoulders (outside diameter) or between umbilical seams (inside diameter).

umbilicus. external depression on each side of shell centered on axis of coiling, its rim being the umbilical shoulder or edge.

undet. an abbreviation for undetermined, meaning that the family, genus or species is not yet determined and thus, unknown.

undulations. a wavy form on the surface that creates swellings and constrictions on the shell.

var. abbreviation for variety or variation, meaning this species is similar to or a variation of a subspecies.

venter. peripheral or exterior siphuncle wall of a whorl comprising the part of the shell that is farthest from the protoconch, generally flat to rounded and bordered by tubercles.

ventral edge. the area where the flanks and venter meet, including the venter.

ventral shoulder. the area on either side of the flank directly bordering the venter.

ventrolateral. on either side of the venter (or keel) along the external edge of the flanks.

whorl. complete turn of shell through 360°.

whorl height. height of whorl measured at right angles to maximum width, the distance from the umbilical seam to the middle of the venter.

whorl section. transverse or **cross section** of a whorl.

whorl width. maximum horizontal distance located between ribs or spines on opposite whorl sides.

zone. a stratigraphic unit in many categories of stratigraphic classification. There are many kinds of zones depending on the stratigraphic properties under consideration (i.e., lithozones, range zones).

Didymoceras cf. *archiacianum*
Drawing by John R. Stacy

BIBLIOGRAPHY

AMMONITES AND THE OTHER CEPHALOPODS OF THE PIERRE SEAWAY

Abbott, R. T., 1974. *American Seashells.* Van Nostrand Reinhold Co., New York, 663 p.

Adkins, W. S., 1929. "Some Upper Cretaceous Taylor ammonites from Texas." *Texas University Bulletin 2901,* pp. 203–211, pls. 5, 6.

Agassiz, L., 1847. "An introduction to the study of natural history, *in* A series of lectures delivered in the hall of the College of Physicians and Surgeons, New York." *Classification of Mollusca, Lecture No. 6,* p. 20.

Agnew, A. F., and Tychsen, P. C., 1965. "A guide to the stratigraphy of South Dakota." *South Dakota State Geological Survey Bulletin 14,* 195 p.

Alexander, L., 1988. "A study in the lineages of the scaphitids of the Western Interior region of the United States." Unpublished manuscript.

Anderson, F. M., and Hanna, G. D., 1935. "Cretaceous geology of Lower California." *Proceedings of the California Academy of Science, Vol. 23(4),* pp. 1–34, pls. 1–11.

Anderson, R. R., and Witzke, B. J., 1994. "The terminal Cretaceous Manson impact structure in north-central Iowa: A window into the Late Cretaceous history of the eastern margin of the Western Cretaceous Seaway," *in* Shurr, G. W., Ludvigson, G. A., and Hammond, R. H., eds., *Perspectives on the eastern margin of the Cretaceous Western Interior Basin.* The Geological Society of America, Boulder, Colorado, *Special Paper 287,* pp. 197–210.

Asquith, D. O., 1970. "Depositional topography and major marine environments, Late Cretaceous, Wyoming." *American Association of Petroleum Geologist Bulletin, Vol. 54(7),* pp. 1184–1224.

Atabekian, A. A., and Khakimov, F. K., 1976. "Campanian and Maastrichtian ammonites of central Asia." *Dushanbe,* pp. 98–99, pl. 11.

Bamburak, J. D., 1978. "Stratigraphy of the Riding Mountain, Boissevain and Turtle Mountain Formations in the Turtle Mountain area, Manitoba." Manitoba Department of Mines, Resources and Environmental Management, *Geological Report 78-2,* 47 p.

Bannatyne, B. B., 1970. "The clays and shales of Manitoba." Manitoba Department of Mines and Natural Resources, *Publication 67-1,* 107 p.

Bandel, K., Landman, N. H., and Waage, K. M., 1982. "Micro-ornament on early whorls of Mesozoic ammonites: Implications for early ontogeny." *Journal of Paleontology, Vol. 56(2),* pp. 386–391.

Bardack, D., 1965. "Anatomy and evolution of chirocentrid fishes." *University of Kansas Paleontological Contributions, Lawrence, Vertebrata, Article 10,* 88 p.

Bardack, D., 1968. "Fossil vertebrates from the marine Cretaceous of Manitoba." *Canadian Journal of Earth Sciences 5,* pp. 145–153.

Bardack, D., and Sprinkle, G., 1969. "Morphology and evolution of saurocephalid fishes." *Fieldiana Geology 16,* pp. 297–340.

Bauer, C. M., 1916. "Contributions to the geology and paleontology of San Juan County, New Mexico; Part 1— Stratigraphy of a part of the Chaco River Valley." *U.S. Geological Survey Professional Paper 98-p,* pp. 271–278.

Beede, J. W., 1900. *Paleontology, Carboniferous and Cretaceous, Vol. VI(II).* The University Geological Survey of Kansas, Lawrence, 516 p.

Bergstresser, T. J., 1983. "Radiolaria from the Upper Cretaceous Pierre Shale, Colorado, Kansas, Wyoming." *Journal of Paleontology, Vol. 57*, pp. 877–882.

Bergstresser, T. J., and Krebs, W. N., 1983. "Late Cretaceous (Campanian–Maastrichtian) diatoms from the Pierre Shale, Wyoming, Colorado and Kansas." *Journal of Paleontology, Vol. 57*, pp. 883–891.

Bishop, G. A., 1967. "Biostratigraphic mapping in the Upper Pierre Shale utilizing the cephalopod genus *Baculites,* Cedar Creek Anticline, Montana." Thesis in partial fulfillment of the requirements for Master of Science in Geology, South Dakota School of Mines and Technology, Rapid City, 18 p.

Bishop, G. A., 1973. *"Homolopsis dawsonensis:* A new crab (Crustacea, Decapoda) from the Pierre Shale (Upper Cretaceous, Maastrichtian) of Cedar Creek Anticline, eastern Montana." *Journal of Paleontology, Vol. 47(1)*, pp. 19–20.

Bishop, G. A., 1976. *"Ekalakia lamberti* N. Gen., N. Sp. (Crustacea, Decapoda) from the Upper Cretaceous Pierre Shale of eastern Montana." *Journal of Paleontology, Vol. 50(3)*, pp. 398–401.

Bishop, G. A., 1978a. "Two new crabs, *Sodakus tatankayotankaensis* N. Gen., N. Sp. and *Raninella oaheensis* N. Sp. (Crustacea, Decapoda), from the Upper Cretaceous Pierre Shale of South Dakota." *Journal of Paleontology, Vol. 52(3)*, pp. 608–617.

Bishop, G. A., 1978b. "Fossil Decapods from the Heart Tail Ranch Locality and the Baresch Locality, Butte, County, South Dakota." Unpublished manuscript.

Bishop, G. A., 1982. *"Homolopsis mendryki*: A new fossil crab (Crustacea, Decapoda) from the Late Cretaceous Dakoticancer Assemblage, Pierre Shale (Maastrichtian) of South Dakota." *Journal of Paleontology, Vol. 56(1)*, pp. 221–225.

Bishop, G. A., 1983. "Two new species of crabs, *Notopocorystes* (Eucorystes) *eichornis* and *Zygastrocarcinus griesi* (Decapoda: Brachyura) from the Bearpaw Shale (Campanian) of north-central Montana." *Journal of Paleontology, Vol. 57(5)*, pp. 900–910.

Bishop, G. A., 1985a. "A new crab, *Eomunidopsis cobbani* N. Sp. (Crustacea, Decapoda), from the Pierre Shale (Early Maastrichtian) of Colorado." *Journal of Paleontology, Vol. 59(3)*, pp. 601–604.

Bishop, G. A., 1985b. "Fossil decapod crustaceans from the Gammon Ferruginous Member, Pierre Shale (Early Campanian), Black Hills, South Dakota." *Journal of Paleontology, Vol. 59(3)*, pp. 605–624.

Bishop, G. A., 1986a. "A new crab, *Zygastrocarcinus cardsmithi* (Crustacea, Decapoda), from the Lower Pierre Shale, Southeastern Montana." *Journal of Paleontology, Vol. 60(5)*, pp. 1097–1102.

Bishop, G. A., 1986b. "Occurrence, preservation, and biogeography of the Cretaceous crabs of North America." *Crustacean Biogeography,* Academy of Natural Science of Philadelphia, pp. 111–142.

Bishop, G. A., 1987. *"Dromiopsis kimberlyae,* A new Late Cretaceous crab from the Pierre Shale of South Dakota." *Proceedings of the Biological Society of Washington, 100:1*, pp. 35–39.

Bishop, G. A., 1996. "Fossil crabs from Tepee Buttes: Cretaceous submarine seeps of the Pierre Shale [abs.]." The Geological Society of America, Rocky Mountain Section, Annual Meeting, Rapid City, South Dakota, *Program with Abstracts*, Boulder, Colorado, p. 3.

de Blainville, H. M. D., 1825. "Nautile." *Dictionnaire des Sciences Naturelles, Vol. 34*, pp. 285–296.

Blueford, J. R., 1988. "Radiolarian evidence: Late Cretaceous through Eocene ocean circulation patterns," *in* Hein, J. R., and Obradovich, J., eds. *Siliceous Deposits of the Tethys and Pacific Region, New York.* Springer-Verlag, New York, pp. 19–29.

Boardman, R. S., Cheetham, A. H., and Rowell, A. J. eds., 1987. "Fossil Invertebrates." *Blackwell Scientific Publications, Palo Alto, California*, pp. 327–339.

Böhm, J., 1898. "Ueber Ammonites pedernalis v. Buch." *Zeitschrift der Deutschen Geologischen Gesellschaft, Vol. 59*, pp. 183–201.

Böse, E., 1927. "Cretaceous ammonites from Texas and northern Mexico." *University of Texas Bulletin, No. 2748*, pp. 143–312, pl. 1–18.

Bottjer, D. J., Hickman, C. S., and Ward, P. D., 1985. "Mollusks: Notes for a short course," Broadhead, T. W., ed. *University of Tennessee Department of Geological Sciences Studies in Geology 13.*

Bowen, C. F., 1915. "The stratigraphy of the Montana Group." *U.S. Geological Survey Professional Paper 90-I*, 153 p.

Bowen, C. F., 1918. "Stratigraphy of the Hanna Basin, Wyoming." *U.S. Geological Survey Professional Paper 108-L*, pp. 227–241.

Bozanic, D., 1995. "A brief discussion on the subsuface Cretaceous rocks of the San Juan Basin." *Geology of parts of Paradox, Black Mesa and San Juan Basins, Four Corners Field Conference, Four Corners Geological Society Guidebook*, pp. 89–107.

Bretz, R., Bishop, G., Fox, J., and Dandavata, K., 1981. "Stratigraphy and depositional environments of Lower and Upper Cretaceous strata, southern Black Hills, South Dakota," *in* Rich, F. J., ed. Geology of the Black Hills, South Dakota and Wyoming, 2nd ed. *American Geological Institute*, pp. 1–18.

Brown, R. W., 1954. *Composition of Scientific Words.* Smithsonian Institution Press, Washington, D.C. 882 p.

Caldwell, W. G. E., 1968. "The Late Cretaceous Bearpaw Formation in the south Saskatchewan River Valley." Saskatchewan Research Council, Geology Division, *Report 8*, pp. 1–86.

Caldwell, W. G. E., (Ed), 1975. *The Cretaceous System in the Western Interior of North America.* The Geological Association of Canada *Special Paper No. 13,* Waterloo, Ontario, 666 p.

Caldwell, W. G. E., 1982. "The Cretaceous system in the Williston Basin—a modern appraisal," *in* Christopher, J. E., and Kent, D. M., eds., *Proceedings of the Fourth International Symposium on the Williston Basin.* Saskatchewan Geological Society, pp. 295–312.

Caldwell, W. G. E., North, B. R., Stelck, C. R., and Wall, J. H., 1978. "A foraminiferal zonal scheme for the Cretaceous System in the Interior Plains of Canada," *in* Stelck, C. R., and Chatterton, B. D. E., eds., *Western and Arctic Canadian Biostratigraphy.* Geological Association of Canada *Special Paper 18*, Waterloo, Ontario, pp. 495–575.

Caldwell, W. G. E., Diner, R., Eicher, D. L., Fowler, S. P., North, B. R., Stelck, C. R., and von Holdt, W. L., 1993. "Foraminiferal biostratigraphy of Cretaceous marine cyclothems," *in* Caldwell, W. G. E., and Kauffman, E. G., eds., *Evolution of the Western Interior Basin.* Geological Association of Canada *Special Paper 39*, pp. 477-520.

Carpenter, K., 1983. "Identification guide to the vertebrate fauna of the Pierre Shale." Unpublished manuscript.

Carpenter, K., 1990. "Continuation of the Upper Smoky Hill vertebrate fauna with the Sharon Springs vertebrate fauna (Campanian)." Unpublished manuscript.

Carroll, R. L., 1988. *Vertebrate Paleontology and Evolution.* W. H. Freeman and Company, New York, 698 p.

Cobban, W. A., 1951a. "New species of *Baculites* from the Upper Cretaceous of Montana and South Dakota." *Journal of Paleontology, Vol. 25(6)*, pp. 817–821.

Cobban, W. A., 1951b. "Colorado Shale of central and northwestern Montana and equivalent rocks of Black Hills." *American Association of Petroleum Geologists Bulletin, Vol. 35(10)*, pp. 2170–2198.

Cobban, W. A., 1952a. "Cretaceous rocks on the north flank of the Black Hills Uplift," *in* Sonnenberg, F.P., ed., *Billings Geological Society Guidebook, Third Annual Field Conference, 1952*, pp. 86–88.

Cobban, W. A., 1952b. "A new Upper Cretaceous ammonite genus from Wyoming and Utah." *Journal of Paleontology, Vol. 26(5)*, pp. 758–760.

Cobban, W. A., 1955. "Cretaceous rocks of northwestern Montana." *Billings Geological Society Guidebook, Sixth Annual Field Conference*, pp. 107–119.

Cobban, W. A., 1958a. "Two new species of *Baculites* from the Western Interior Region." *Journal of Paleontology, Vol. 32(4)*, pp. 660–665.

Cobban, W. A., 1958b. "Late Cretaceous fossil zones of the Powder River Basin, Wyoming and Montana." *Wyoming Geological Association Guidebook, Thirteenth Annual Field Conference, Powder River Basin*, pp. 114–119.

Cobban, W. A., 1962a. "New *Baculites* from the Bearpaw Shale and equivalent rocks of the Western Interior." *Journal of Paleontology, Vol. 36(1)*, pp. 126–135.

Cobban, W. A., 1962b. "*Baculites* from the lower part of the Pierre Shale and equivalent rocks in the Western Interior." *Journal of Paleontology, Vol. 36(4)*, pp. 704–718.

Cobban, W. A., 1963. "Occurrence of the Late Cretaceous ammonite, *Hoplitoplacenticeras*, in Wyoming." *U.S. Geological Survey Professional Paper 475-C, Part C.*

Cobban, W. A., 1964. "The Late Cretaceous cephalopod *Haresiceras* Reeside and its possible origin." *U.S. Geological Survey Professional Paper 454I*, 19 p., 3 pls.

Cobban, W. A., 1969. "The Late Cretaceous ammonites *Scaphites leei* Reeside and *Scaphites hippocrepis* (DeKay) in the Western Interior of the United States." *U.S. Geological Survey Professional Paper 619.*

Cobban, W. A., 1970. "Occurrence of the Late Cretaceous ammonites *Didymoceras stevensoni* (Whitfield) and *Exiteloceras jenneyi* (Whitfield) in Delaware," *in Geological Survey Research 1970, Chapter D., U.S. Geological Survey Professional Paper 700-D*, pp. D71–D76.

Cobban, W. A., 1973. "The Late Cretaceous Ammonite *Baculites undatus* Stephenson in Colorado and New Mexico." *U.S. Geological Survey Journal of Research, Vol. 1(4)*, pp 459–465.

Cobban, W. A., 1974a. "Some ammonoids from the Ripley Formation of Mississippi, Alabama, and Georgia." *U.S. Geological Survey Journal of Research, Vol. 2(1)*, pp. 81–88, 6 figs.

Cobban, W. A., 1974b. "Ammonites from the Navesink Formation at Atlantic Highlands, New Jersey." *U.S. Geological Survey Professional Paper 845*, 21p., 11 pls.

Cobban, W. A., 1976. "Ammonite Record from the Pierre Shale of northeastern New Mexico." *New Mexico Geological Society 27th Field Conference, Vermejo Park*, pp. 165–169.

Cobban, W. A., 1977. "A new curved baculite from the Upper Cretaceous of Wyoming." *U.S. Geological Survey Journal of Research, Vol. 5(4)*, pp. 457–462.

Cobban, W. A., 1987. "The Upper Cretaceous ammonite *Rhaeboceras* Meek in the Western Interior of the United States." *U.S. Geological Survey Professional Paper 1477.*

Cobban, W. A., 1993. "Diversity and distribution of Late Cretaceous ammonites, Western Interior, United States," *in* Caldwell, W. G. E., and Kaufman, E. G., eds., *Evolution of the Western Interior Basin. Geological Association of Canada Special Paper 39, Waterloo, Ontario*, pp. 435–451.

Cobban, W. A., Hook, S. D., and Kennedy, W. J., 1989. "Upper Cretaceous rocks and ammonite faunas of southwestern New Mexico." *New Mexico Bureau of Mines and Mineral Resources Memoir 45, Socorro,* 137 p.

Cobban W. A., and Jeletzky, J. A., 1965. "A new scaphite from the Campanian rocks of the Western Interior of North America." *Journal of Paleontology, Vol. 39(5),* pp. 794–801.

Cobban, W. A., and Kennedy, W. J., 1991a. "Some Upper Cretaceous ammonites from the Nacatoch Sand of Hempstead County, Arkansas." *U.S. Geological Survey Bulletin 1985-C,* pp. C1–C5.

Cobban, W. A., and Kennedy, W. J., 1991b. "New records of the ammonite subfamily *Texanitinae* in Campanian (Upper Cretaceous) rocks in the Western Interior of the United States." *U.S. Geological Survey Bulletin 1985, Part B.*

Cobban, W. A., and Kennedy, W. J., 1991c. "*Pachydiscus* (Ammonoidea) from Campanian (Upper Cretaceous) rocks in the Western Interior of the United States." *U.S. Geological Survey Bulletin 1985, Part F.*

Cobban, W. A., and Kennedy, W. J., 1993a. "The Upper Cretaceous dimorphic pachydiscid ammonite *Menuites* in the Western Interior of the United States." *U.S. Geological Survey Professional Paper 1533,* 14 p.

Cobban, W. A., and Kennedy, W. J., 1993b. "Middle Campanian ammonites and inoceramids from the Wolfe City Sand in northeastern Texas." *Journal of Paleontology 67(1),* pp. 71–82.

Cobban, W. A., and Kennedy, W. J., 1994a. "Upper Cretaceous ammonites from the Coon Creek Tongue of the Ripley Formation at its Type locality in McNairy County, Tennessee." *U.S. Geological Survey Bulletin 2073-B,* 15 p., 11 pls.

Cobban, W. A., and Kennedy, W. J., 1994b. "A giant baculite from the Upper Campanian and Lower Maastrichtian of the Western Interior." *U.S. Geological Survey Bulletin 2073-C,* pp. C1–C3, pl. 2.

Cobban, W. A., and Kennedy, W. J., 1994c. "Middle Campanian (Upper Cretaceous) ammonites from the Pecan Gap Chalk of central and northeastern Texas." *U.S. Geological Survey Bulletin 2073-D,* pp. D1–D9, pl. 5.

Cobban, W. A., and Kennedy, W. J., 1995. "Maastrichtian ammonites chiefly from the Prairie Bluff Chalk in Alabama and Mississippi." *Journal of Paleontology Memoir 44, Vol. 69, part III of III, Supplement to No. 5,* 40 p.

Cobban, W. A., Kennedy, W. J., and Scott, G. R., 1992. "Upper Cretaceous heteromorph ammonites from the *Baculites compressus* Zone of the Pierre Shale in north central Colorado." *U.S. Geological Survey Bulletin 2024, Part A.*

Cobban, W. A., Kennedy, W. J., and Scott, G. R., In press. "Some upper Campanian (Upper Cretaceous) heteromorph ammonites from the Western Interior of the United States." *U.S. Geological Survey Professional Paper.*

Cobban, W. A., Landis, E. R., and Dane, C. H., 1974. "Age relations of upper part of Lewis Shale on east side of San Juan Basin, New Mexico." *Geological Society Guidebook, Twenty-Fifth Field Conference, Ghost Ranch (central northern New Mexico),* pp. 279–282.

Cobban, W. A., and Larson, N. L., In press. "Marine Upper Cretaceous rocks and their ammonite record along the northern flank of the Black Hills Uplift, Montana, Wyoming, and South Dakota." South Dakota School of Mines & Technology, Museum of Geology, Rapid City, *Dakotera.*

Cobban, W. A., Merewether, E. A., Fouch, T. D., and Obradovich, J. D., 1994. "Some Cretaceous shorelines in the Western Interior of the United States." *Mesozoic Systems of the Rocky Mountain Region, USA.*

Cobban, W. A., and Reeside, J. B., Jr., 1952. "Correlation of the Cretaceous formations of the Western Interior of the United States." *The Geological Society of America Bulletin, Vol. 63, Boulder, Colorado,* pp. 1011–1044.

Cobban, W. A., and Scott, G. R., 1964. "Multinodose scaphitid cephalopods from the lower part of the Pierre Shale and equivalent rocks in the conterminous United States." *U.S. Geological Survey Professional Paper 483-E,* 13 p.

Cobban, W. A., Scott, G. R., and Gill, J. R., 1962. "Recent discoveries of the Cretaceous ammonite *Haresiceras* and their stratigraphic significance." *U.S. Geological Survey Professional Paper 450-B*, pp. 58–60.

Collignon, M., 1948. "Ammonites néocrétacées de Menabe (Madagascar), I, Les Texanitidae." *Annales géologiques de Service des Mines, No. 13*, pp. 49–104, pls. 7–20; *No. 14*, pp. 5–101, pls. 15–32.

Collignon, M., 1969. "Atlas des fossiles caractértistiques de Madagascar (Ammonites), pt. 15, Campanien inférieur." *Republique Malgache Service Géologique*, 216 p., pls. 514–606.

Collier, A. J., 1918. "Geology of northeastern Montana." *U.S. Geological Survey Professional Paper 120-B*, pp. 17–39.

Conrad, T. A., 1857. "Descriptions of Cretaceous and Tertiary Fossils." *Emory, Report on the United States and Mexican Boundary Survey, Vol. 1.*

Conrad, T. A., 1860. "Descriptions of new species of Cretaceous and Eocene fossils of Mississippi and Alabama." *Journal of the Academy of Natural Science of Philadelphia, 2d Ser., Vol. 4*, pp. 275–298.

Conrad, T. A., 1868. "Synopsis of the invertebrate fossils of the Cretaceous Formation of New Jersey," *in* Cook, J. H. ed., *Geology of New Jersey. Geological Survey of New Jersey*, pp. 721–732.

Conrad, T. A., 1874. "Descriptions of new mollusks from Cretaceous beds of Colorado." *U.S. Geological and Geographical Survey of the Territories Annual Report 7*, pp. 455–456.

Cook, T. D., and Bally, A. W., eds, 1975. *Stratigraphic Atlas of North and Central America.* Princeton University Press, Princeton, New Jersey, 272 p.

Cooke, C. W., 1953. "American Upper Cretaceous Echinoidea." *U.S. Geological Survey Professional Paper 254-A*, 44 p., 16 pls.

Cope, E. D., 1875. "The vertebrata of the Cretaceous Formations of the West." *U.S. Geological Survey of the Territories, Vol. II*, 302 p.

Coquand, H., 1859. "Synopsis des animaux et des végétaux fossiles observés dans la formation crétacée du Sud-Ouest de la France." *Bull. Soc. Géol. Fr., Vol. 16(2)*, pp. 945–1023.

Coryell, H. N., and Salmon, E. S., 1934. "A molluscan faunule from the Pierre Formation in eastern Montana." *American Museum Novitates, Vol. 746*, pp. 1–18.

Crandall, D., 1958. "Geology of the Pierre Area South Dakota." *U.S. Geological Survey Professional Paper 307*, pp. 1–83.

Cuvier, G., 1797. *Tableau élémentaire de l'histoire naturelle des animaux.* Paris, 710 p.

Dane, C. H., Pierce, W. G., and Reeside, J. B., 1936. "The stratigraphy of the Upper Cretaceous rocks north of the Arkansas River in eastern Colorado." *U.S. Geological Survey Professional Paper 186-K*, pp. 224–232.

Dante, J. H., 1942. "Description of fossil fishes from the Upper Cretaceous of North America." *American Journal of Science 240*, pp. 339–348.

Darby, D. G., and Ojakangas, R. W., 1980. "Gastroliths from an Upper Cretaceous plesiosaur." *Journal of Paleontology 54*, pp. 548–556.

Darton, N. H., 1900. "Geology and water resources of the southern half of the Black Hills and adjoining regions in South Dakota and Wyoming." *U.S. Geological Survey Twenty First Annual Report*, pp. 489–599.

Darton, N. H., 1902. "Description of the Oelrichs Quadrangle, South Dakota-Nebraska." *U.S. Geological Survey, Geologic Atlas, Oelrichs Folio 85*, 6 p.

Darton, N. H., 1904a. "Comparison of the stratigraphy of the Black Hills, Bighorn Mountains and Rocky Mountain Front Range." *The Geological Society of America Bulletin, Vol. 15, Boulder, Colorado,* pp. 379–448, pls. 23–26.

Darton, N. H., 1904b. "Description of the Newcastle Quadrangle, Wyoming–South Dakota." *U.S. Geological Survey, Geologic Atlas, Newcastle Folio 107,* 9 p.

Darton, N. H., 1908. "Paleozoic and Mesozoic of central Wyoming." *The Geological Society of America Bulletin, Vol. 19, Boulder, Colorado,* pp. 403–470, pls. 21–30.

Darton, N. H., 1909. "Geology and water resources of the northern portion of the Black Hills and adjoining regions in South Dakota and Wyoming." *U.S. Geological Survey Professional Paper 65,* 105 p.

Darton, N. H., 1919. "Description of the Newell Quadrangle, South Dakota." *U.S. Geological Survey, Geologic Atlas, Newell Folio 209,* 7 p.

Darton, N. H., and O'Harra, C. C., 1905. "Description of the Aladdin Quadrangle, Wyoming–South Dakota–Montana." *U.S. Geological Survey, Geologic Atlas, Aladdin Folio 128,* 8 p.

Darton, N. H., and O'Harra, C. C., 1909. "Description of the Belle Fourche Quadrangle, South Dakota." *U.S. Geological Survey, Geologic Atlas, Belle Fourche Folio 164,* 9 p.

Darton, N. H., and Smith, W. S. T., 1904. "Description of the Edgemont Quadrangle, South Dakota–Nebraska." *U.S. Geological Survey, Geologic Atlas, Edgemont Folio 108,* 10 p.

DeKay, J. E., 1827. "Report on several multilocular shells from the State of Delaware; With observations of a second specimen of the new fossil genus *Eurypterus.*" *Annals of the Lyceum of Natural History of New York, Vol. 2,* pp. 273–279.

DeWitt, E., Redden, J. A., Buscher, D., and Wilson, A. B., 1989. "Geologic map of the Black Hills area, South Dakota and Wyoming." *U. S. Geological Survey Miscellaneous Investigations Series Map I-1910, scale 1:250,000.*

Denson, N. M., Gibson, M. L., and Sims, G. L., 1993. "Geologic map showing thickness of the Upper Cretaceous Pierre Shale in the northern half of the Powder River Basin, southeastern Montana and northeastern Wyoming." *U.S. Geological Survey Miscellaneous Investigations Series Map I-2380-A, scale 1:200,000.*

Derstler, K., 1985. "Pierre mosasaur expedition." *University of New Orleans Field Report,* 60 p.

Donnan, K., 1994. "Garmon and siphuncle." *Lapidary Journal, July Issue,* pp. 18–23.

Douglas, E., 1909. "A geological reconnaissance in North Dakota, Montana, and Idaho; with notes on Mesozoic and Cenozoic geology." *Annals of the Carnegie Museum, Vol. V(2&3),* pp. 211–288.

Douglas, R. J. W., 1942. "New species of *Inoceramus* from the Cretaceous Bearpaw Formation." *Transactions of The Royal Society of Canada, Third Series, Vol. IIIVI(IV),* pp. 59–68.

Dowling, D. B., 1917. "The southern plains of Alberta." *Canadian Department of Mines Geological Survey, Memoir 93, Geological Series No. 78,* 200 p.

Dunkle, D. H., 1960. "Three North American Cretaceous fishes." *Proceedings of the United States National Museum, Vol. 108,* pp. 269–277.

Dyman, T. S., Cobban, W. A., Fox, J. E., Hammond, R. H., Nichols, D. J., Perry, W. J., Jr., Porter, K. W., Rice, D. D., Setterholm, D. R., Shurr, G. W., Tysdall, R. G., Haley, J. C., and Campen, E. B., 1994. "Cretaceous rocks from southwestern Minnesota, northern Rocky Mountains, and Great Plains region," *in* Shurr, G. W., Ludvigson, G. A., and Hammond, R. H., eds., *Perspectives on the Eastern margin of the Cretaceous Western Interior Basin. The Geological Society of America Special Paper 287, Boulder, Colorado,* pp. 5–26.

Dyman, T. S., Merewether, E. A., Molenaar, C. M., Cobban, W. A., Obradovich, J. D., Weimer, R. J., and Bryant, W. A., 1994. "Stratigraphic transects for Cretaceous rocks, Rocky Mountains and Great Plains region," *in* Caputo, M. V., Peterson, J. A., and Franczyk, K. J., eds., *Mesozoic Systems of the Rocky Mountain Region, USA.* The Rocky Mountain Section of the Society for Sedimentary Geology, pp. 365–392.

Eaton, J. G., 1991. "Biostratigraphic framework for the Upper Cretaceous rocks of the Kaiparowits Plateau, southern Utah." *The Geological Society of America Special Paper 260, Boulder, Colorado,* pp. 47–64.

Eaton, J. G., and Nations, J. D., 1991. "Introduction; Tectonic setting along the margin of the Cretaceous Western Interior Seaway, southwestern Utah and northern Arizona." *The Geological Society of America Special Paper 260, Boulder, Colorado,* pp. 1–8.

Elias, M. K., 1933. "Cephalopods of the Pierre Formation of Wallace County, Kansas, and adjacent area." *University of Kansas Science Bulletin, Vol. 21(9), Lawrence,* pp. 289–363.

Ellis, M. S., and Colton, R. B., 1994. "Geologic map of the Powder River Basin and surrounding area, Wyoming, Montana, South Dakota, North Dakota, and Nebraska." *U.S. Geological Survey Miscellaneous Investigations Series Map I-2298, scale 1:500,000.*

Evans, J., and Shumard, B. F., 1854. "Descriptions of new fossil species from the Cretaceous Formation of Sage Creek, Nebraska." *Philadelphia Academy of Natural Science Proceedings, Vol. 7,* pp. 163–164.

Fassett, J. E., 1987. "The ages of the continental, Upper Cretaceous, Fruitland Formation and Kirtland Shale based on a projection of ammonite zones from the Lewis Shale, San Juan Basin, New Mexico and Colorado." *Geological Society of America Special Paper 209, Boulder, Colorado,* pp. 5–16.

Fassett, J. E., and Rigby, K. J., Jr., eds., 1987. *The Cretaceous–Tertiary Boundary in the San Juan and Raton Basins, New Mexico and Colorado. The Geological Society of America Special Paper 209, Boulder, Colorado,* 200 p.

Feldman, R. M., Bishop, G. A., and Kammer, T. W., 1977. "Macrurous decapods from the Bearpaw Shale (Cretaceous, Campanian) of northeastern Montana." *Journal of Paleontology 51,* pp. 1161–1180, 3 pls.

Fisher, D. J., Erdmann, C. E., and Reeside, J. B., Jr., 1960. "Cretaceous and Tertiary formations of the Book Cliffs, Carbon, Emery, and Grand Counties, Utah; and Garfield and Mesa Counties, Colorado." *U.S. Geological Survey Professional Paper 332,* 80 p., 12 pls.

Flores, R. M., Hohman, J. C., and Ethridge, F. G., 1991. "Heterogeneity of Upper Cretaceous Gallup Sandstone regressive facies, Gallup Sag, New Mexico." *The Geological Society of America Special Paper 260, Boulder, Colorado,* pp. 189–210.

Fox, J. E., 1996. "The manganese-bearing Degrey Member of the Pierre Shale, an inner to middle shelf facies associated with the Precambrian Sioux Ridge, central South Dakota [abs.]." The Geological Society of America, Rocky Mountain Section, Annual Meeting, Rapid City, South Dakota, *Program with Abstracts,* p. 8.

Gardiner, B., 1966. "Catalogue of Canadian fossil fishes." *Royal Ontario Museum Life Sciences Contribution 68,* 154 p.

Giebel, C. G., 1850. "Spezielle darstellung der gattung *Scaphites*." *Jahresbericht des Naturwissen schaftlichen vereins Halle,* pp. 18–20.

Gill, J. R., and Cobban, W. A., 1962. "Red Bird Silty Member of the Pierre Shale, a new stratigraphic unit." *U.S. Geological Survey Professional Paper 450B,* pp. B21–B24.

Gill, J. R., and Cobban, W. A., 1965. "Stratigraphy of the Pierre Shale, Valley City and Pembina Mountain Areas, North Dakota." *U.S. Geological Survey Professional Paper 392-A,* 20 p.

Gill, J. R., and Cobban, W. A., 1966a. "The Red Bird Section of the Upper Cretaceous Pierre Shale in Wyoming." *U.S. Geological Survey Professional Paper 393-A,* 73 p.

Gill, J. R., and Cobban, W. A., 1966b. "Regional unconformity in Late Cretaceous, Wyoming." *U.S. Geological Survey Professional Paper 550-B*, pp. 20–27.

Gill, J. R., Cobban, W. A., and Schultz, L. G., 1972a. "Stratigraphy and composition of the Sharon Springs Member of the Pierre Shale in western Kansas." *U.S. Geological Survey Professional Paper 728*, 50 p.

Gill, J. R., Cobban, W. A., and Schultz, L. G., 1972b. "Correlation, ammonite zonation, and a reference section for the Montana Group, central Montana, in Crazy Mountains Basin, 1972." *Montana Geological Society, Twenty First Field Conference Guidebook*, pp. 91–97.

Gill, J. R., and Cobban, W. A., 1973. "Stratigraphy and geologic history of the Montana Group and equivalent rocks, Montana, Wyoming, and North and South Dakota." *U.S. Geological Survey Professional Paper 776*, 37 p.

Gill, J. R., Merewether, E. A., and Cobban, W. A., 1970. "Stratigraphy and nomenclature of some Upper Cretaceous and Lower Tertiary rocks in south-central Wyoming." *U.S. Geological Survey Professional Paper 667*, 53 p.

Gill, T., 1871. "Arrangement of the families of mollusks." *Smithsonian Miscellaneous Collections 227*, 49 p.

Given, M. M., and Wall, J. H., 1971. "Microfauna from the Upper Cretaceous Bearpaw Formation of north-central Alberta." *Bulletin of Canadian Petroleum Geology, Vol. 19*, pp. 504–546.

Goody, P. C., 1970. "The Cretaceous teleostian fish *Cimolichthys* from the Niobrara Formation of Kansas and Pierre Formation of Wyoming." *American Museum of Natural History Novitates 2434*, 29 p.

Goody, P. C., 1976. "*Enchodus* (Teleostei: Enchodontidae) from the Upper Cretaceous Pierre Shale of Wyoming and South Dakota with an evaluation of the North American enchodontid species." *Palaeontographica Abt. A. Bd 152: 4-6*, pp. 93–112.

Gordon, W. A., 1973. "Marine life and ocean surface currents in the Cretaceous." *Journal of Geology, Vol. 81*, pp. 269–284.

Green, W. D., 1964. "Greybull Giant." *The Billings Gazette, Morning Edition, Montana, 5 July*, p. 7.

Gregory, H. E., and Moore, R. C. "The Kaiparowits Region — a geographic and geologic reconnaisance of parts of Utah and Arizona." *U.S. Geological Survey Professional Paper 164*, 157 p.

Grier, J. C., Grier, J. W., and Peterson, J. G., 1993. "Occurrence of the Upper Cretaceous ammonite *Rhaeboceras* in the *Baculites eliasi* Zone of the Pierre Shale." *Journal of Paleontology 66(3)*, pp. 521–523.

Gries, J. P., 1942. "Economic possibilities of the Pierre Shale." *South Dakota Geological Survey Report of Investigations 43*, 79 p.

Gries, J. P., 1953. "Cretaceous rocks of Williston Basin." *American Association of Petroleum Geologists Bulletin, Vol. 38*, pp. 443–453.

Gries, J. P., and Rothrock, E. P., 1941. "Manganese deposits of the lower Missouri Valley in South Dakota." *South Dakota Geological Survey Report of Investigations 38*, 96 p.

Hall J., and Meek, F. B., 1854. *Memoir of the Academy of Arts and Sciences, Boston V (n.s.)*, 396 p.

Hall J., and Meek, F. B., 1856. "Descriptions of new species of fossils from the Cretaceous formations of Nebraska, with observations upon *Baculites ovatus* and *Baculites compressus*, and the progressive development of the septa in baculites, ammonites, and scaphites." *American Academy of Arts and Science Memoir, new ser., Vol. 5*, pp. 379–411.

Hancock, J. M., and Kauffman, E. G., 1989. "Use of eustatic changes of sea level to fix Campanian–Maastrichtian boundary in Western Interior of USA [abs.]." *Twenty-Eighth International Geological Congress, Abstracts, Vol. 2(23)*.

Hanczaryk, P. A., Miller, K. G., Gallagher, W. B., Feigenson, M. D., and Van Fossen, M. C., 1996. "Stratigraphy of the Campanian-Maastrichtian Pierre Shale of central South Dakota [abs.]." The Geological Society of America, Boulder, Colorado, Rocky Mountain Section, Annual Meeting, Rapid City, South Dakota, *Program with Abstracts*, p. 10

Hattin, D. E., and Cobban, W. A., 1965. "Upper Cretaceous stratigraphy, paleontology and paleoecology of western Kansas." *The Geological Survey of America, Kansas City,* 66 p.

Haun, J. D., and Weimer, R. J., 1961. "Cretaceous stratigraphy of Colorado." *Guide to the Geology of Colorado*, pp. 58–65.

Hayden, F. V., 1859. "Catalogue of the collections in geology and natural history obtained by the expedition under command of Lieutenant G. K. Warren," *in* Warren, G. K., ed., *Preliminary Account of Explorations in Nebraska and Dakota in the Years 1855–56–57. U.S. War Department Annual Report 1858 (U.S. Congress, Second Session. House Executive Document 2)*, pp. 673–705.

Heald, K. C., 1926. "The geology of the Ingomar Anticline Treasure and Rosebud counties, Montana." *U.S. Geological Survey Bulletin 786-A*, 37 p.

Henderson, J., 1904. "Paleontology of the Boulder Area." *The University of Colorado Studies, Vol. II(2)*, pp. 95–106.

Henderson, J., 1908. "New species of Cretaceous invertebrates from northern Colorado." *U.S. National Museum Proceedings, Vol. XXXIV*, pp. 259–264.

Henderson, J., 1920. "The Cretaceous Formations of northeastern Colorado and the Foothills Formations of north-central Colorado." *Colorado Geological Survey Bulletin 19*, 57 p.

Hintze, L. F., 1973. "Geologic history of Utah." *Brigham Young University Geology Studies, Vol. 20(3)*, 181 p.

Hirsch, K. F., 1975. "Die Ammoniten des Pierre Meeres (Oberkreide) in den westlichen USA." *Der Aufschluss, Jahrgang 26*, pp. 103–113.

Hoganson, J. W., Hanson, M., Halvorson, D. L., and Halvorson, V. G., 1996. "Mosasaur remains and associated vertebrate fossils from the Degrey Member (Campanian) of the Pierre Shale, Cooperstown Site, Griggs County, east central North Dakota [abs.]." The Geological Society of America, Boulder, Colorado, Rocky Mountain Section, Annual Meeting, Rapid City, South Dakota, *Program with Abstracts*, p. 11.

Horner, J. R., 1979. "Upper Cretaceous dinosaurs from the Bearpaw Shale (marine) of south-central Montana with a checklist of Upper Cretaceous dinosaur remains from marine sediments in North America." *Journal of Paleontology 53*, pp. 566–577.

House, M. R., and Senior, J. R., eds, 1981. *The Ammonoidea (The Evolution, Classification, Mode of Life, and Geological Usefulness of a Major Fossil Group)*. The Systemetics Association, *Special Vol. 18*, Academic Press, New York, 593 p.

Hutchinson, P. J., and Kues, B. S., 1985. "Depositional environments and paleontology of Lewis Shale to Lower Kirtland Shale sequence (Upper Cretaceous) Bisti area, northwestern New Mexico." *New Mexico Bureau of Mines and Mineral Resources Circular 195, Socorro*, pp. 25–54.

Hyatt, A., 1894. "Phylogeny of an acquired characteristic." *Proceedings of the American Philosophical Society, Vol. 32(143)*, pp. 349–647, pls. 1–14.

Hyatt, A., 1900. "Cephalopoda," *in* Zittel, K. A. [1896-1900]. *Textbook of Paleontology*. Macmillan, London, pp. 502–604.

Hyatt, A., 1903. "Pseudoceratites of the Cretaceous." *U.S. Geological Survey Monographs, Vol. XLIV*, 351 p.

Izett, G. A., Cobban, W. A., and Gill, J. R., 1971. "The Pierre Shale near Kremmling, Colorado, and its correlation to the east and the west." *U.S. Geological Survey Professional Paper 684-A*, 19 p., 1 pl.

Izett, G. A., Cobban, W. A., Obradovich, J. D., and Kunk, M. J., 1993. "The Manson Impact structure: ^{40}Ar/^{39}Ar Age and its distal impact ejects in the Pierre Shale in southeastern South Dakota." *Science, Vol. 262*, pp. 729–732.

Jagt, J. W. M., and Kennedy, W. J., 1994. "*Jeletzkytes dorfi* Landman and Waage 1993, a North American ammonoid marker from the lower Upper Maastrichtian of Belgium, and the numerical age of the Lower/Upper Maastrichtian boundary." *N. Jb. Geol. Paläont. Mb., Vol. 11(4)*, pp. 239–245.

Jeletzky, J. A., 1955. "Evolution of Santonian and Campanian *Belemnitella* and paleontological systematics: exemplified by *Belemnitella praecursor* Stolley." *Journal of Paleontology, Vol. 29(3)*, pp. 478–509, pls. 56–58.

Jeletzky, J. A., 1960. "Youngest marine rocks in Western Interior of North America and the age of the *Triceratops* beds, with remarks on comparable dinosaur-bearing beds outside North America." Twenty-first International Geological Congress, Copenhagen, *Report, pt. 5*, pp. 25–40.

Jeletzky, J. A., 1962. "Study of *Scaphites* fauna from the Bearpaw Formation of Alberta and Saskatchewan." *Geological Survey of Canada Paper 62-3, 4 p.*

Jeletzky, J. A., 1966. "Mollusca." *The University of Kansas Paleontological Contributions Article 7, Lawrence,* 162 p.

Jeletzky, J. A., 1968. "Macrofossil zones of the marine Cretaceous of the Western Interior of Canada and their correlation with the zones and stages of Europe and the Western Interior of the United States." *Geological Survey of Canada Paper*, pp. 67–72.

Jeletzky, J. A., 1970. "Cretaceous macrofaunas," *in Geology and Economic Minerals of Canada, Part B, Economic Geology Report 1*, pp. 649–662.

Jeletzky, J. A., 1971. "Cretaceous biotic provinces and paleogeography of western and Arctic Canada: Illustrated by a detailed study of ammonites." *Geological Survey of Canada Paper 70-22*, pp. 1–92.

Jensen, F. S., and Varnes, H. D., 1964. "Geology of the Fort Peck Area, Garfield, McCone and Valley Counties, Montana." *U.S. Geological Survey Professional Paper 414-F*, 49 p.

Jerzykiewicz, T., 1996. "*Baculites compressus robinsoni* Cobban from the Crowsnest River section at Lundbreck, Alberta: an implication for the timing of the late Cretaceous Bearpaw transgression into the southern Foothills of Alberta," *in Current Research 1996-E*; Geological Survey of Canada Project 930006, Calgary, Alberta, pp. 97-100.

Jerzykiewicz, T., Sweet, A. R., and McNeil, D. H., 1996. "Shoreface of the Bearpaw Sea in the footwall of the Lewis Thrust, southern Canadian Cordillera, Alberta," *in Current Research 1996-E,* Geological Survey of Canada Project 930006, Calgary, Alberta, pp. 155-163.

Johnson, R. C., and Keighin, C. W., 1981. "Cretaceous and Tertiary History and Resources of the Piceance." *The New Mexico Geological Society, Inc., 32nd Annual Field Conference, Western Slope, Colorado and Utah, Guidebook*, pp. 199–210.

Jones, D. L., 1963. "Upper Cretaceous (Campanian and Maastrichtian) ammonites from southern Alaska." *U.S. Geological Survey Professional Paper 432*, 53 p., 41 pl.

Jones, R. L., 1979. "Mineral dispersal patterns in the Pierre Shale." Ph.D. thesis, University of Oklahoma, Norman, 270 p.

Jones, R. L., and Blatt, H., 1984. "Mineral dispersal patterns in Pierre Shale." *Journal of Sedimentary Petrology, Vol. 54*, p. 17–28.

Jorgensen, S., 1991. "Endemic faunas, locations of measurable sections and the approximate thicknesses of the biostratigraphic zones in the Campanian Age Upper Cretaceous Pierre Shale in a portion of Fall River County, South Dakota." Unpublished map.

Kammer, T. W., and Raff, R. A., 1978. "Additional information on *Hoploparia bearpawensis* Feldman (Crustacea: Decapoda) from the Bearpaw Shale (Cretaceous; Campanian) of northeastern Montana." *Journal of Paleontology 52*, pp. 1338–1339.

Kauffman, E. G., 1967. "Cretaceous *Thyasira* from the Western Interior of North America." *Smithsonian Miscellaneous Collections, Vol. 152(1)*, 159 p.

Kauffman, E. G., 1975. "Dispersal and biostratigraphic potential of Cretaceous benthonic bivalvia in the Western Interior," *in* Caldwell, W. G. E., ed., *The Cretaceous System in the Western Interior of North America. Geological Association of Canada Special Paper 13, Waterloo, Ontario*, pp. 163–194.

Kauffman, E. G., ed, 1977. "Cretaceous facies, faunas and paleoenvironments across the Western Interior Basin." *The Mountain Geologist Field Guide, Vol. 14(3 & 4)*.

Kauffman, E. G., 1984. "Paleobiogeography and evolutionary response dynamic in the Cretaceous Western Interior Seaway of North America," *in* Westerman, G. E. G., ed., *Jurassic-Cretaceous Biochronology and Paleo-geography of North America. Geological Association of Canada Special Paper 27, Waterloo, Ontario*, pp. 273–306.

Kauffman, E. G., and Caldwell, W. G. E., 1993. "The Western Interior Basin in space and time," *in* Caldwell, W. G. E., and Kauffman, E. G., eds., *Evolution of the Western Interior Basin. Geological Association of Canada Special Paper 39*, pp. 1-30.

Kauffman, E. G., and Kesling, R. V., 1960. "An Upper Cretaceous ammonite bitten by a mosasaur." *University of Michigan Contributions from the Museum of Paleontology 15*, pp. 193–248.

Kauffman, E. G., Sageman, B. B., Kirkland, J. L., Elder, W. P., Harries, P. J., and Villamil, T., 1993. "Molluscan beostratigraphy of the Cretaceous Western Interior Basin, North America," *in* Caldwell, W. G. E., and Kauffman, E. G., eds., *Evolution of the Western Interior Basin. Geological Association of Canada Special Paper 39*, pp. 397-434.

Keefer, W. R., 1965. "Stratigraphy and geologic history of the uppermost Cretaceous, Paleocene, and Lower Eocene rocks in the Wind River Basin, Wyoming." *U.S. Geological Survey Professional Paper 495-A*, 77 p.

Keefer, W. R., 1972. "Frontier, Cody, and Mesaverde Formations in the Wind River and southern Bighorn Basins, Wyoming." *U.S. Geological Survey Professional Paper 495-E*, 23 p.

Kennedy, W. J., and Cobban, W. A., 1976. "Aspects of ammonite biology, biogeography, and biostratigraphy." *The Palaeontological Association Special Papers in Palaeontology No. 17*, 94 p.

Kennedy, W. J., and Cobban, W. A., 1992. "Ammonite correlation of the uppermost Campanian of western Europe, the U.S. Gulf Coast, Atlantic seaboard, and Western Interior, and the numerical age of the base of the Maastrichtian." *Geological Magazine, Vol. 129*, pp. 497–500.

Kennedy, W. J., and Cobban, W. A., 1993a. "Upper Campanian ammonites from the Ozan–Annona Formation boundary in southwestern Arkansas." *Bulletin Geological Society of Denmark, Vol. 40*, pp. 115–148.

Kennedy, W. J., and Cobban, W. A., 1993b. "Campanian ammonites from the Annona Chalk near Yancy, Arkansas." *Journal of Paleontology, Vol. 67(1)*, pp. 83–97.

Kennedy, W. J., and Cobban, W. A., 1993c. "Ammonites from the Saratoga Chalk (Upper Cretaceous) Arkansas." *Journal of Paleontology, Vol. 67(3)*, pp. 404–434.

Kennedy, W. J., Cobban, W. A., and Landman, N. H., 1996. "Two species of *Placenticeras* (Ammonitina) from the Upper Cretaceous (Campanian) of the Western Interior of the United States." *American Museum of Natural History Novitates.*

Kennedy, W. J., Cobban, W. A., and Landman, N. H., In press. "The Maastrichtian ammonites *Coahuilites sheltoni* Böse (1927), and *Sphenodiscus pleurisepta* (Conrad, 1857), from the uppermost Pierre Shale and basal Fox Hills Formation of Colorado and Wyoming." *American Museum of Natural History Novitates,* 7 p.

Kennedy, W. J., Cobban, W. A., and Scott, G. R., 1992. "Ammonite correlation of the uppermost Campanian of Western Europe, the U. S. Gulf Coast, Atlantic Seaboard and Western Interior, and the numerical age of the base of the Maastrichtian." *Geological Magazine 129(4),* pp. 497–500.

Kennedy, W. J., Johnson, R. O., and Cobban, W. A., 1995. "Upper Cretaceous ammonite faunas of New Jersey." *Proceedings of a Geological Association of New Jersey Symposium, Contributions to the Paleontology of New Jersey, Vol XII,* pp. 24–55.

Kirk, S. R., 1930. "Cretaceous stratigraphy of the Manitoba Escarpment." *Geological Survey of Canada, Summary Report, 1929, Part B,* pp. 112B–135B.

Kitchell, J. A., and Clark, D. A., 1982. "Late Cretaceous-Paleogene paleogeography and paleocirculation: Evidence of north polar upwelling." *Palæogeography, Palæoclimatology, Palæoecology, Vol. 40,* pp. 135–165.

Klassen, R. W., Wyder, J. R., and Bannatyne, B. B., 1970. "Bedrock topography and geology of southern Manitoba." *Geological Survey of Canada, Paper 70-15 (map 25-1970, scale 1:500,000).*

Knechtel, M. M., and Patterson, S. H., 1955. "Bentonite deposits of the northern Black Hills District, Montana, Wyoming, and South Dakota." *U.S. Geological Survey Mineral Investigations Field Studies Map MF 36, 2 sheets, scale 1:48,000, with sections and text.*

Knechtel, M. M., and Patterson, S. H., 1962. "Bentonite deposits of the northern Black Hills District, Wyoming, Montana, and South Dakota." *U.S. Geological Survey Bulletin 1082-M,* pp. 893–1030.

Koeberl, C., and Anderson, R. R., 1996. "The Manson Impact Structure, Iowa: Anatomy of an impact crater." *The Geological Society of America Special Paper 302, Boulder, Colorado,* 468 p.

Kullman, J., and Wiedmann, J., 1970. "Significance of sutures in phylogeny of Ammonoidea." *University of Kansas Paleontological Contributions Paper, Lawrence, 47,* 32 p.

Kummel, B., 1956. "Post-Triassic nautiloid genera." *Bulletin of the Museum of Comparative Zoology, Vol. 114(7).*

Lamarck, J. B. P. A. de M. de., 1799. "Prodrome d'une nouvelle classification des coquilles." *Mémoires de la Société d'Histoire Naturelle de Paris (for 1799),* pp. 63–90.

Landes, R. W., 1940. "Paleontology of the marine formations of the Montana group. Pt. 2 of Geology of the southern Alberta plains." *Geological Survey of Canada Memoir 221,* pp. 129–217, 8 pls.

Landman, N. H., 1982. "Embryonic shells of *Baculites*." *Journal of Paleontology, Vol. 56(5),* pp. 1235–1241.

Landman, N. H., 1985. "Preserved ammonitellas of *Scaphites* (Ammonoidea, Ancyloceratina)." *American Museum Novitates No. 2815,* pp. 1–10.

Landman, N. H., 1988. "Heterochrony in ammonites." *in* McKinney, M. L., ed., *Heterochrony in Evolution.* Plenum Publishing Corporation, New York, pp. 159–182.

Landman, N. H., and Bandel, K., 1985. "Internal structures in the early whorls of Mesozoic ammonites." *American Museum Novitates, No. 2823,* pp. 1–21.

Landman, N. H., Cochran, J. K., Rye, D. M., Tanabe, K., and Arnold, J. M., 1994. "Early life history of *Nautilus*: Evidence from isotopic analysis of aquarium-reared specimens." *Paleobiology, Vol. 20(1)*, pp. 40–51.

Landman, N. H., Kennedy, W. J., Cobban, W. A., and Larson, N. L., 1996. "Scaphitid ammonoids from the Upper Cretaceous Pierre Shale [abs.]." The Geological Society of America, Boulder, Colorado, Rocky Mountain Section, Annual Meeting, Rapid City, South Dakota, *Program with Abstracts*, p. 14.

Landman, N. H., Rye, D., and Shelton, K., 1983. "Early ontogeny of *Eutrephoceras* compared to recent *Nautilus* and Mesozoic ammonites: Evidence from shell morphology and light stable isotopes." *Paleobiology, Vol. 9*, pp. 269–279.

Landman, N. H., and Waage, K. M., 1993. "Scaphitid ammonites of the Upper Cretaceous (Maastrichtian) Fox Hills Formation in South Dakota and Wyoming." *American Museum of Natural History Bulletin No. 215*.

Larson, N. L., 1992. "Some common ammonites of the Pierre Shale." Unpublished manuscript.

Larson, P. L., 1989. *What Is an Ammonite?* Black Hills Institute of Geological Research, Inc., Hill City, South Dakota, pamphlet and poster.

Larson, P. L., Larson, N. L., and Farrar, R. A., 1991. *The Pierre Shale and Its Macrofauna.* Black Hills Institute of Geological Research, Inc. publication, Hill City, South Dakota, pp. 1–17.

Lee, W. T., 1915. "Relation of the Cretaceous formations to the Rocky Mountains in Colorado and New Mexico." *U.S. Geological Survey Professional Paper 95-C*, pp. 27–58, 5 pls.

Lee, W. T., 1927. "Correlation of geological formations between east-central Colorado, Central Wyoming and southern Montana." *U.S. Geological Survey Professional Paper 149*, 88 p., 35 pls.

Lehmann, U., 1966. "Dimorphismus bei Ammoniten der Ahrensburger Lias-Geschiebe." *Paläontologie Z., Vol. 40*, pp. 26–55.

Lehmann, U., 1967. "Ammoniten mit Kieferapparat und radula aus Lias-Geschieben." *Paläontologie Z., Vol. 41*, pp. 38–45.

Lehmann, U., 1981. *The Ammonites: Their Life and their World.* Cambridge University Press, New York, 246 p.

Lillegraven, J. A., and Ostresh, L. M., 1990. "Late Cretaceous (earliest Campanian/Maastrichtian) evolution of western shorelines of the North American Western Interior Seaway in relation to known mammalian faunas," *in* Bown, T. M., and Rose, K. D., eds., *Dawn of the Age of Mammals in the Northern Part of the Rocky Mountain Interior, North America.* The Geological Society of America, Boulder, Colorado *Special Paper 243*.

Lines, F. G., 1963. "Stratigraphy of Bearpaw Formation of southern Alberta." *Bulletin of Canadian Petroleum Geology, Vol. 11(3)*, pp. 212–227.

Link, T. A., and Childerhose, A. J., 1931. "Bearpaw Shale and contiguous formations in Lethbridge area, Alberta." *American Association of Petroleum Geologists Bulletin, Vol. 15(10)*, pp. 1227–1242.

Lisenbee, A. L., 1985. "Tectonic map of the Black Hills Uplift, Montana, Wyoming, and South Dakota." *Geological Survey of Wyoming Map Series 13*.

Logan, W. N., 1898. "The invertebrates of the Benton, Niobrara, and Fort Pierre Groups." *Kansas University Geological Survey, Vol. 4(8)*, pp. 431–518, pls. 86–120.

Loranger, D. M., and Gleddie, J., 1953. "Some Bearpaw Zones in southwestern Alberta." *Proceedings of the Third Annual Field Conference and Symposium, Alberta Society of Petroleum Geologists*, pp. 158–175.

Love, J. D., and Weitz, J. L., 1951. "Geological map of the Powder River Basin and adjacent areas, Wyoming." *U.S. Geological Survey Oil and Gas Investigations Map OM 122, scale 1 in. to 5 mi.*

Lucas, S. G., Reser, P. K., and Wolberg, D. L., 1985. "Shark vertebrae from the Upper Cretaceous, Pierre Shale, northeastern New Mexico." *New Mexico Bureau of Mines and Mineral Resources Circular 195, Socorro,* pp. 21–24.

McNeil, D. H., 1984. "The eastern facies of the Cretaceous System in the Canadian Western Interior," *in* Stott, D. F., and Glass, D. J., eds., *The Mesozoic of Middle North America, Calgary, Alberta. Canadian Society of Petroleum Geologists, Memoir 9,* pp. 145–171.

McNeil, D. H., and Caldwell, W. G. E., 1981. "Cretaceous rocks and their foraminifera in the Manitoba escarpment." *Geological Association of Canada Special Paper 21, Waterloo, Ontario,* 439 p.

Mallory, W. W., ed., 1972. *Geologic Atlas of the Rocky Mountain Region U.S.A.* Rocky Mountain Association of Geologists, A. B. Hirschfield Press, Denver, Colorado, 331 p.

Maloney, C., 1996. "Biostratigraphy of Late Cretaceous marine Chelonia of North America [abs.]." The Geological Society of America, Boulder, Colorado, Rocky Mountain Section, Annual Meeting, Rapid City, South Dakota, *Program with Abstracts,* p. 16.

Mapel, W. J., Robinson, C. S., and Theobald, P. K., 1959. "Geologic and structure contour map of the northern and western flanks of the Black Hills, Wyoming, Montana, and South Dakota." *U.S. Geological Survey Oil and Gas Investigations Map OM-191, 2 sheets, scale 1:96,000, with text.*

Marsh, O. C., 1880. "Odontornithes, monograph on the extinct toothed birds of North America." *U.S. Geological Exploration of the Fortieth Parallel Report,* 201 p.

Martin, J. E., 1996. "Disconformities of the Lower Pierre Shale (Cretaceous) of South Dakota [abs.]." The Geological Society of America, Boulder, Colorado, Rocky Mountain Section, Annual Meeting, Rapid City, South Dakota, *Program with Abstracts,* p. 16.

Martin, J. E., Bell, G. L., and Bertog, J. L., 1996. "Biostratigraphic ranges and biozonation of mosasaurs (Reptilia) within the Late Cretaceous of the North American epicontinental seaway [abs.]." The Geological Society of America, Boulder, Colorado, Rocky Mountain Section, Annual Meeting, Rapid City, South Dakota, *Program with Abstracts,* p. 16.

Martin, J. E., Bell, G. L., Jr., Shumacher, B. A., and Sawyer, J. F., 1996. "Geology and paleontology of Late Cretaceous marine deposits of the southern Black Hills region: Road log, Field Trip 8." *in* Paterson, C. J., and Kirchner, J. G., eds., *Guidebook to the Geology of the Black Hills, South Dakota.* South Dakota School of Mines and Technology, Department of Geology and Geological Engineering, Rapid City, *Bulletin No. 19,* pp. 51–77.

Martin, J. E., and Bjork, P. R., 1987. "Gastric residues associated with a mosasaur from the Late Cretaceous (Campanian) Pierre Shale in South Dakota." *in* Martin, J. E., and Ostrander, G. E., eds., *Papers in Vertebrate Paleontology in Honor of Morton Green.* South Dakota School of Mines and Technology, Museum of Geology, Rapid City, Dakotera, *Vol. 3,* pp. 68–72.

Martin, L. D., and Stewart, J. D., 1982. "An ichthyornithiform bird from the Campanian of Canada." *Canadian Journal of Earth Sciences, Vol. 19,* pp. 324–327.

Matsumoto, T., 1867. "Evolution of the Nostoceratidae (Cretaceous heteromorph ammonoids)." *Memoir Fac. Sci., Kyushu Univ., Serv. D, Geology, Vol. KVIII(2),* pp. 331–347, pls. 18–19.

Meek, F. B., 1860. "Systematic catalogue, with synonyma, etc., of Jurassic, Cretaceous and Tertiary fossils collected in Nebraska, by the exploring expeditions under the command of Lieutenant G. K. Warren, of the U. S. Topographical Engineers." *Proceedings of the Academy of Natural Science of Philadelphia, Vol. 12,* pp. 417–432.

Meek, F. B., 1862. "Descriptions of new Cretaceous fossils from Nebraska Territory, collected by the expedition sent out by the Government under the command of Lieutenant John Mullan, U.S. Topographical Engineers, for the location and construction of a wagon road from the sources of the Missouri to the Pacific Ocean." *Academy of Natural Science of Philadelphia Proceedings 1862,* pp. 21–28.

Meek, F. B., 1870. "A preliminary list of fossils collected by Dr. Hayden in Colorado, New Mexico, and California." *American Philosophical Society Proceedings, Vol. II,* 429 p.

Meek, F. B., 1872. "Hayden's second annual report." *U.S. Geological Survey of the Territories,* 297 p.

Meek, F. B., 1876. "A report of invertebrate Cretaceous and Tertiary fossils of the Upper Missouri country F. V. Hayden in charge, Reports." *U.S. Geological Survey of the Territories, Vol. IX,* 629 p., pls. 1–45.

Meek, F. B., and Hayden, F. V., 1856. "Descriptions of new species of gastropoda and cephalopoda from the Cretaceous formations of Nebraska Territory." *Proceedings of the Academy of Natural Science of Philadelphia, Vol. 8,* pp. 63–126.

Meek, F. B., and Hayden, F. V., 1857. "*Ptychoceras mortoni.*" *Proceedings of the Academy of Natural Science of Philadelphia, Vol. IX,* 134 p.

Meek, F. B., and Hayden, F. V., 1858. "Descriptions of new organic remains collected in Nebraska Territory, together with some remarks on the geology of the Black Hills and portions of the surrounding country." *Proceedings of the Academy of Natural Science of Philadelphia, Vol. 8,* pp. 41–59.

Meek, F. B., and Hayden, F. V., 1860. "Descriptions of new organic remains from the Tertiary, Cretaceous, and Jurassic rocks of Nebraska." *Proceedings of the Academy of Natural Science of Philadelphia, Vol. 12,* pp. 175–185.

Meek, F. B., and Hayden, F. V., 1861. "Descriptions of new Lower Silurian (Primordial), Jurassic, Cretaceous and Tertiary fossils collected in Nebraska by the exploring expedition under the command of Captain W. F. Raynolds, U.S. Topographical Engineer, with some remarks on the rocks from which they were obtained." *Philadelphia Academy of Natural Science Proceedings,* pp. 415–447.

Meek, F. B., and Hayden, F. V., 1862. "*Nautilus elegans* var. *Nebrascensis.*" *Proceedings of the Academy of Natural Science in Philadelphia, Vol. XIV.*

Mello, J. E., 1969. "Foraminifera and stratigraphy of the upper part of the Pierre Shale and lower part of the Fox Hills Sandstone (Cretaceous) north-central South Dakota." *U.S. Geological Survey Professional Paper 611,* pp. 1–121, 12 pls.

Mickelson, J. C., and Bishop, G. A., 1966. "An occurrence of the young of *Baculites compressus* Say Ss., Meade County, South Dakota." *Proceedings of the South Dakota Academy of Science Vol. 45(22),* pp. 62–66.

Mikhailov, N. P., 1951. "[The ammonites of the Upper Cretaceous of the southern part of the European part of the U.S.S.R.], Inst. geol. nauk Trudy, no. 129." *Geological Series No. 50,* 143 p.

Miller, A. K., and Youngquist, W., 1946. "A giant ammonite from the Cretaceous of Montana." *Journal of Paleontology, Vol. 20(5),* pp. 73–75.

Miller, H. W., Jr., 1957. "*Niobrarateuthis bonneri,* a new genus and species of squid from the Niobrara Formation of Kansas." *Journal of Paleontology, Vol, 31(4),* pp. 809–814.

Molenaar, C. M., and Rice, D. D., 1988. "Cretaceous rocks of the Western Interior Basin." *in* Sloss, L L., ed., *Sedimentary Cover—North American Craton: Boulder, Colorado.* The Geological Society of America, Boulder, Colorado, *The Geology of North America, Vol. D-2,* pp. 77–82.

Momper, J. A., and Tyrrell, W. W., Jr., 1957. "Catalog of stratigraphic names of the southwest San Juan Basin and adjacent areas." *Geology of southwestern San Juan Basin, Second Four Corners Field Conference, Four Corners Geological Society Guidebook,* pp. 17–24.

Moore, R. C., 1920. "Oil and gas resources of Kansas—Part II Geology of Kansas." *State Geological Survey of Kansas, Lawrence, Bulletin 6.*

Moore, R. C., eds., 1957. *Treatise on Invertebrate Paleontology, Part L, Mollusca 4.* The Geological Society of America, Boulder, Colorado, and University of Kansas Press, Lawrence.

Moore, R. C., eds., 1964. *Treatise on Invertebrate Paleontology, Part K, Mollusca 3.* The Geological Society of America, Boulder, Colorado, and University of Kansas Press, Lawrence.

Morrison, J., and Brand, U., 1986. "Paleooceanography of the upper-mid-Cretaceous Seaway of Canada [abs]." Geological Association of Canada Annual Meeting, Saskatoon, Saskatchewan, *Program with Abstracts*, p. 104.

Morrison, J., and Brand, U., 1987. "Chemostratigraphic evaluation of Upper Cretaceous molluscs from the Interior Seaway, western Canada [abs]." Geological Association of Canada Annual Meeting, Saskatoon, Saskatchewan, *Program with Abstracts*, p 75.

Morton, S. G., 1830. "Synopsis of the organic remains of the Ferruginous Sand Formation of the United States, with geological remarks." *American Journal of Science, 1st ser., Vol., 18*, pp. 243–250.

Morton, S. G., 1834. *Synopsis of the Organic Remains of the Cretaceous Group of the United States.* W. P. Gibbons, Philadelphia, 88 p., 19 pls., app. 1–8.

Morton, S. G., 1841. "Description of several new species of fossil shells from the Cretaceous deposits of the United States." *Proceedings of the Academy of Natural Sciences of Philadelphia, Vol.1*, pp. 106–110.

Morton, S. G., 1842. "Description of some new species of organic remains of the Cretaceous Group of the United States with a tabular view of the fossils hitherto discovered in this formation." *Journal of the Academy of Natural Science of Philadelphia, Vol. 8(2)*, pp. 207–227.

Murray, D. K., 1981. "Upper Cretaceous (Campanian) coal resources of Western Colorado." *The New Mexico Geological Society, Inc., 32nd Annual Field Conference, Western Slope, Colorado and Utah, Guidebook*, pp. 233–240.

Nace, R. L., 1936. "Summary of the Late Cretaceous and Early Tertiary stratigraphy of Wyoming." *The Geological Survey of Wyoming Bulletin No. 26*, 271 p.

Naef, A., 1916. "Systematische Übersicht der mediterranan Cephalopoden." *Stazione Zool. Napoli*, pp. 11–19.

Naef, A., 1921. "Das system der dibranchiaten cephalopoden und die mediterranen Arten derselben." *Mitteilungen aus der zool. Station zu Seapel, Vol. 22*, pp. 527–542.

Nascimbene, G. G., 1965. "Bentonites and the geochronology of the Bearpaw Sea." *Bulletin of the Canadian Petroleum Geology, Vol. 13*, 537 p.

Nations, J. D., and Eaton, J. G., 1991. "Stratigraphy, depositional environments and sedimentary tectonics of the western margin, Cretaceous Western Interior Seaway." *The Geological Society of America Special Paper 260, Boulder, Colorado*, 211 p.

Nelson, G. E., ed., 1985. *The Cretaceous Geology of Wyoming. Wyoming Geological Association 36th Annual Field Conference Guidebook*, 184 p.

Newman, K. R., 1987. "Biostratigraphic correlation of Cretaceous-Tertiary boundary rocks, Colorado to San Juan Basin, New Mexico." *The Geological Society of America Special Paper 209, Boulder, Colorado*, pp. 151–164.

Nicholls, E. L., 1987. "The vertebrate fauna of the Pembina Member of the Pierre Shale (Lower Campanian) of southern Manitoba." *Journal of Vertebrate Paleontology, Vol. 7 (supplement to number 3)*, pp. 21–22.

Nicholls, E. L., 1988. "New material of *Toxochelys latiremis* Cope, and a revision of the Genus *Toxochelys* (Testudines, Chelonioidea)." *Journal of Vertebrate Paleontology, Vol. 8(2)*, pp. 181–187.

Nicholls, E. L., and Isaak, H., 1987. "Stratigraphic and taxonomic significance of *Tusoteuthis longa* Logan (*Coleoidea, Teuthida*) from the Pembina Member, Pierre Shale (Campanian, of Manitoba)." *Journal of Paleontology, Vol. 61(4)*, pp. 727–737.

Nichols, T. C., Jr., Chleborad, A. F., and Collins, D. S., 1987. "Government Draw Bentonite Beds—A newly identified stratigraphic marker in the Virgin Creek Member of the Pierre Shale, central South Dakota." *The Mountain Geologist, Vol. 24(3)*, pp. 77–80.

Nichols, T. C., Jr., and Collins, D. S., 1986. "*In situ* and laboratory geotechnical tests in the Pierre Shale near Hayes, South Dakota." *U.S. Geological Survey Open File Report 86-152*, 24 p.

Nichols, T. C., Jr., Collins, D. S., and Davidson, R. R., 1986. "*In situ* and laboratory geotechnical tests in the Pierre Shale near Hayes, South Dakota—a characterization of engineering behavior." *Canadian Geotechnical Journal, Vol. 23*, pp. 181–194.

Nichols, T. C., Jr., Collins, D. S., Jones-Cecil, M., and Swolfs, H. S., 1994. "Faults and structure in the Pierre Shale, central South Dakota," *in* Shurr, G. W., Ludvigson, G. A., and Hammond, R. H., eds., *Perspectives on the Eastern Margin of the Cretaceous Western Interior Basin.* The Geological Society of America *Special Paper 287*, Boulder, Colorado, pp. 211–236.

Nichols, T. C., Jr., King, K. W., Collins, D. S., and Williams, R. A., 1998. "Seismic-reflection technique used to verify shallow rebound fracture zones in the Pierre Shale of South Dakota." *Canadian Geotechnical Journal, Vol. 25*, pp. 369–374.

Norton, O. R., 1994. *Rocks from Space: Metetorites and Meteorite Hunters.* Mountain Press Publishing, Missoula, Montana, 446 p.

Nowak, T., 1911. "Untersuchungen über die Cephalopoden der oberen Kreide in Polen, II Teil: Die Skaphiten." *Bulletin de L'Academie des Sciences de Cracovie, Série, B*, pp. 547 580, pls. XXXII XXXIII.

Nowak, T., 1913. "Untersuchungen über die Cephalopoden der oberen Kreide in Polen, III Teil: Anzeiger der wissenschaften in Krakau, Mathematisch-Naturwissenschaftliche Klasse." *Reihe B: Biologische wissenschaften, Année 1913*, pp. 335–415, pls. 40–45, 1 fig.

Obradovich, J. D., 1993. "A Cretaceous time scale," *in* Caldwell, W. G. E., and Kauffman, E. G., eds., *Evolution of the Western Interior Basin. Geological Association Canada Special Paper 39, Waterloo, Ontario*, pp. 379–396.

Obradovich, J. D., and Cobban, W. A., 1975. "A time-scale for the Late Cretaceous of the Western Interior of North America," *in* Caldwell, W. G. E., ed., *The Cretaceous System in the Western Interior of North America. Geological Association of Canada Special Paper 13, Waterloo, Ontario*, pp. 31–54.

d'Orbigny, A., 1840–1849. *Paléontologie française. Térains jurassiques, vol. 1, Céphalopodes (1842–1849). Térrains crétacés. vol. 1, Céphalopoes (1840–1842).* Author's press (Paris).

d'Orbigny, A., 1850–1852. "Prodome de paléontologie stratigraphique universelle des animaux mollusques et rayonnés." *Vol. 2, V. Masson, Paris*, 428 p.

Orlov, Y. A., Luppov, N. P., and Drushchits, V. V., eds., 1976. *Fundamentals of Paleontology, Vol. VI, Mollusca-Cephalopoda II, Ammonoidea (Ceratitidae and Ammonitidae), Endocochlia, Coniconchia.* Translated from Russian, Israel Program for Scientific Translations, Jerusalem, 474 p.

Osterwald, F. W., and Dean, B. G., 1957. "Preliminary tectonic map of western South Dakota, showing the distribution of uranium deposits: U. S. Geological Survey Mineral." *Investigations Field Studies Map MF 128, scale 1:500,000.*

Osterwald, F. W., and Dean, B. G., 1958. "Preliminary tectonic map of eastern Montana showing the distribution of uranium deposits: U. S. Geological Survey Mineral." *Investigations Field Studies Map MF 126, scale 1:500,000.*

Owen, D. D., 1852. *Report of a Geological Survey of Wisconsin, Iowa, and Minnesota and Incidentally of a Portion of Nebraska Territory.* Lippincott, Grambo & Co., Philadelphia, Pennsylvania, 638 p.

Palmer, A. R., 1983. *The Decade of North American Geology 1983 Geologic Time Scale.* The Geological Society of America, Boulder, Colorado.

Parkinson, J., 1811. *Organic Remains of Former Worlds, Vol 3.* Sherwood, Neely & Jones, London, 479 p.

Parrish, J. T., Gaynor, G. D., and Swift, D. J. P., 1984. "Circulation in the Cretaceous Western Interior Seaway of North America, a review," *in* Stott, D. F., and Glass, D. J., eds., *The Mesozoic of Middle North America: Calgary, Alberta.* Canadian Society of Petroleum Geologists, *Memoir 9*, pp. 221–231.

Pavlow, A. P., 1913. "Les céphalopodes du Jura el du crétacé inférieur de la Sibérie septentrionale." *Nouv. Mém. Soc. Imp. Nat. Mus., Moscow, Ser. 8*, pp. 1–68, pl. 8.

Pierce, W. G., and Hunt, C. B., 1937. "Geology and mineral resources of north-central Chouteau, western Hill, and eastern Liberty Counties, Montana." *U.S. Geological Survey Bulletin 847-F*, pp. 225–270.

Pillmore, C. L., and Flores, R. M., 1987. "Stratigraphy and depositional environments of the Cretaceous–Tertiary boundary clay and associated rocks, Raton Basin, New Mexico and Colorado." *The Geological Society of America Special Paper 209, Boulder, Colorado*, pp. 111–130.

Porter, K. W., Campen, E. B., Cobban, W. A., and Hansen, W. B., 1990. "Stratigraphic sequences and unconformities in Cretaceous rocks of east-central Montana [abs.]." *American Association of Petroleum Geologists Bulletin, Vol. 74(8)*, p. 1341.

Rathbun, M. J., 1917. "New species of South Dakota Cretaceous crabs." *Proceedings of the U.S. National Museum, Vol. 52(2182)*, pp. 385–391.

Reese, V. R., 1957. "Cretaceous oil and gas horizons of the San Juan Basin, Colorado and New Mexico." Geology of southwestern San Juan Basin, Second Four Corners Field Conference, *Four Corners Geological Society Guidebook*, pp. 36–39.

Reeside, J. B., Jr., 1927. "The cephalopods of the Eagle Sandstone and related formations in the Western Interior of the United States." *U.S. Geological Survey Professional Paper 151*, 87 p.

Reynolds, M. W., 1966. "Stratigraphic relations of Upper Cretaceous rocks, Lamont-Bairoil area, south-central Wyoming." *U.S. Geological Survey Professional Paper 328*, 124 p.

Riccardi, A. C., 1983. "Scaphitids from the Upper Campanian-Lower Maastrichtian Bearpaw Formation of the Western Interior of Canada." *Geological Survey of Canada Bulletin 354.* 51 p.

Rice, D. D., and Shurr, G. W., 1983. "Patterns of sedimentation and paleogeography across the Western Interior seaway during time of deposition of Upper Cretaceous Eagle Sandstone and equivalent rocks, northern Great Plains," *in* Reynolds, M. W., and Dolly, E. D., eds., *Mesozoic Paleogeography of the West-Central United States.* Society of Economic Paleontologists and Mineralogists Symposium 2, pp. 337–358.

Rich, F. J., (Ed), 1981. "Geology of the Black Hills, South Dakota and Wyoming." *American Geological Institute Field Trip Guidebook for Annual Meeting, Rapid City, South Dakota*, 292 p.

Robinson, C. S., Mapel, W. J., and Bergendahl, M. H., 1964. "Stratigraphy and structure of the northern and western flanks of the Black Hills Uplift, Wyoming, Montana, and South Dakota." *U.S. Geological Survey Professional Paper 404*, 134 p.

Robinson, C. S., Mapel, W. J., and Cobban, W. A., 1959. "Pierre Shale along western and northern flanks of Black Hills, Wyoming and Montana." *American Association of Petroleum Geologists Bulletin, Vol. 43(1)*, pp. 101–123.

Robinson, H. R., 1945. "New *Baculites* from the Cretaceous Bearpaw Formation of southeastern Saskatchewan." *Royal Society of Canada Transactions, ser. 3, Vol. 39(4)*, pp. 51–54, 1 pl.

Roemer, C. F., 1849. *Mir besonderer Rücksicht auf deutsche Auswanderung und die physischen Verhältnisse des Landes nach eigener Beobachtung geschildert; mit einem naturwissenschaftlichen Anhänge, und einer topographisch-geognostischen Karte van Texas.* Bonn, 464 p.

Roemer, F. A., 1841. *Die Versteinerungen des Norddeutschen Kredegebirges.* Hanover, Germany, 145 p., 16 pls.

Roemer, F. A., 1852. *Die Kreidebildungen von Texas und ihre organischen organischen Enschlüsse.* Adolph Marcus. Bonn vi + 100p., 11pl.

Rogers, R. R., 1996. "A nonmarine perspective on the sedimentology of the Claggett and Bearpaw Transgressions [abs.]." The Geological Society of America, Boulder, Colorado, Rocky Mountain Section, Annual Meeting, Rapid City, South Dakota, *Program with Abstracts*, p. 30.

Rosene, R. K., 1972. "Micropaleontology of the Bearpaw Formation, southwestern Alberta Foothills." Thesis in partial fulfillment of the requirements for Master of Science from the University of Alberta, Edmonton, Alberta, 132 p.

Rubey, W. W., 1930. "Lithologic studies of fine-grained Upper Cretaceous sedimentary rocks of the Black Hills region." *U.S. Geological Survey Professional Paper 165-A*, pp. 1–54.

Russell, D. A., 1967. "Systematics and morphology of American mosasaurs (Reptilia, Sauria)." *Peabody Museum of Natural History, Yale University Bulletin 23*, 241 p.

Russell, D. A., 1975. "A new species of *Globidens* from South Dakota, and a review of the globidentine mosasaurs." *Fieldiana Geology 33*, pp. 235–256.

Russell, D. A., 1988. "A check list of North American marine Cretaceous vertebrates including fresh water fishes." *Royal Tyrrell Museum of Palæontology Occasional Paper No. 4*, 57 p.

Russell, L. S., 1970. "Correlation of the Upper Cretaceous Montana Group between southern Alberta and Montana." *Canadian Journal of Earth Sciences, Vol. 7*, pp. 1099-1108.

Russell, L. S., and Landes, R. W., 1940. "Geology of the southern Alberta plains." Department of Mines and Resources, Canada, *Geological Survey Memoir 221*.

Salvador, A., ed., 1994. *International Stratigraphic Guide. A guide to stratigraphic classification, terminology, and procedure, 2nd ed.* The International Union of Geological Sciences and The Geological Society of America, Boulder, Colorado, 214 p.

Saunders, W. B., and Landman, N. H., eds., 1987. *Nautilus. The Biology and Paleobiology of a Living Fossil. Topics in Geobiology, Vol. 6,* Plenum Press, New York.

Say, T., 1820. "Observations on some species of zoophytes, shells, etc., principally fossil." *American Journal of Science, 1st ser., Vol. 2*, pp. 34–45.

Schlüter, C., 1872. "Cephalopoden der oberen deutschen Kreide." *Palæontographica Vol. 21*, pp. 1–264.

Schultz, L. G., 1965. "Mineralogy and stratigraphy of the lower part of the Pierre Shale, South Dakota and Nebraska." *U.S. Geological Survey Professional Paper 1064-A*, 28 p.

Schultz, L. G., 1978. "Mixed-layer clay in the Pierre Shale and equivalent rocks, northern Great Plains region." *U.S. Geological Survey Professional Paper 1064-A*, 28 p.

Schultz, L. G., Tourtelot, H. A., Gill, J. R., and Boerngen, J. G., 1980. "Composition and properties of the Pierre Shale and equivalent rocks, northern Great Plains region." *U.S. Geological Survey Professional Paper 1064-B*, 114 p.

Schultze, H. P., Hunt, J. C, and Neuner, A. M., 1985. "Type and figured specimens of fossil vertebrates in the collection of the University of Kansas Museum of Natural History, Part II Fossil amphibians and reptiles." *University of Kansas Museum of Natural History Miscellaneous Publication 77*, 53 p.

Scott, G. R., and Cobban, W. A., 1959. "So-called Hygiene groups of northeastern Colorado," *in Rocky Mountain Association of Geologists Guidebook.* Eleventh Annual Field Conference Symposium on Cretaceous Rocks of Colorado and Adjacent Areas, pp. 124–131.

Scott, G. R., and Cobban, W. A., 1963. "Apache Creek Sandstone Member of the Pierre Shale of southeastern Colorado." *U.S. Geological Survey Professional Paper 475B, Article 25*, pp. B99–B101.

Scott, G. R., and Cobban, W. A., 1965. "Geologic and biostratigraphic map of the Pierre Shale between Jarre Creek and Loveland, Colorado." *U.S. Geological Survey Miscellaneous Geologic Investigations map I-439, scale 1:48,000, separate text.*

Scott, G. R., and Cobban, W. A., 1975. "Geologic and biostratigraphic map of the Pierre Shale in the Canon City-Florence basin and the Twelvemile Park area, south-central Colorado." *U.S. Geological Survey Miscellaneous Geologic Investigations map I-937, scale 1:48,000.*

Scott, G. R., and Cobban, W. A., 1986a. "Geologic and biostratigraphic map of the Pierre Shale in the Colorado Springs–Pueblo area, Colorado." *U.S. Geological Survey Miscellaneous Geological Investigations map I-1627, scale 1:100,000.*

Scott, G. R., and Cobban, W. A., 1986b. "Geologic, biostratigraphic and structure map of the Pierre Shale between Loveland and Round Butte, Colorado." *U.S. Geological Survey Miscellaneaous Geologic Investigations map I-1700, scale 1:50,000.*

Scott, W. B., 1907. *An Introduction to Geology, 2nd edition.* Macmillan, New York, 816 p.

Searight, W. V., 1937. "Lithologic stratigraphy of the Pierre Formation in the Missouri valley in South Dakota." *South Dakota Geological Survey Report Inventory 27*, 63 p.

Sears, J. D., Hunt, C. B., and Hendricks, T. A., 1938–1940. "Transgressive and regressive Cretaceous deposits in southern San Juan Basin, New Mexico." *U.S. Geological Survey Professional Paper 193-F*, p. 101–121, pls. 25–31.

Sepkoski, J. J., Jr., 1992. *A Compendium of Fossil Marine Animal Families,* 2nd ed. *Milwaukee Public Museum Contributions in Biology and Geology, No. 83*, 156 p.

Sharpe, D., 1853-1856. "Descriptions of the fossil remains of Mollusca found in the Chalk of England." *Monograph Palæontographical Society (London), Pt. 1*, pp 1–26, *and Pt. 2*, pp. 36–70.

Shumard, B. F., 1861. "Descriptions of new Cretaceous fossils from Texas." *Boston Society of Natural History Proceedings, Vol. 8*, pp. 188–205.

Shurr, G. W., Hammond, R. H., and Bretz, R. F., 1994. "Cretaceous paleotectonism and postdepositional tectonism in south-central South Dakota: An example of epeirogenic tectonism in continental lithosphere," *in* Shurr, G. W., Ludvigson, G. A., and Hammond, R. H., eds., *Perspectives on the Eastern Margin of the Cretaceous Western Interior Basin.* The Geological Society of America *Special. Paper 287, Boulder, Colorado*, pp. 237–256.

Shurr, G. W., Ludvigson, G. A., and Hammond, R. H., eds., 1994. "Perspectives on the eastern margin of the Cretaceous Western Interior Basin." *The Geological Society of America Special Paper 287, Boulder, Colorado*, 264 p.

Silver, C., 1957. "Relation of Coastal and submarine topography to Cretaceous stratigraphy." Geology of southwestern San Juan Basin, Second Four Corners Field Conference, *Four Corners Geological Society Guidebook*, pp. 128–137.

Simpson, H. E., 1960. "Geology of the Yankton area, South Dakota and Nebraska." *U.S. Geological Survey Professional Paper 328*, 124 p.

Smith, A. G., Smith, D. G., and Funnell, B. M., 1994. *Atlas of Mesozoic and Cenozoic Coastlines.* Cambridge University Press, pp. 1–99.

Smith, J. H., 1961. "A summary of stratigraphy and paleontology, upper Colorado and Montanan Groups, south-central Wyoming, northeastern Utah and northwestern Colorado." *16th Annual Field Conference, Wyoming Geological Association Guidebook*, pp. 101–112.

Sohl, N. F., 1967. "Upper Cretaceous gastropods from the Pierre Shale at Red Bird, Wyoming." *U.S. Geological Survey Professional Paper 393-B*, 46 p.

Sowerby, (1816). Min. Conch., II, 33, pl. 116 - Mantell (1822), Geol. Sussex, 112, tab. xx, fig. 1. - Sharpe (1853), Monogr. "Chalk Ceph. of England." *Palæontogr. Soc.*, 12 pl. iii, fig. 3; and pl. iv, fig. 1.

Spath, L. F., 1922. "On the Senonian ammonite fauna of Pondoland." *Transactions of the Royal Society of South Africa, Vol. 10(3)*, pp. 113–147, pls. 5–9.

Spath, L. F., 1925. "On Senonian Ammonoidea from Jamaica." *Geological Magazine, Vol. 62*, pp. 28–32.

Spath, L. F., 1926. "On new ammonites from the English Chalk." *Geological Magazine, Vol. 63(740)*, pp. 77–83.

Spivey, R. C., 1940. "Bentonite in southwestern South Dakota." *South Dakota State Geological Survey Report Inventory 36*, 56 p., 7 pls.

Stafford, P. W., and Chamberlain, J. A., Jr., 1996. "The Upper Pierre Shale/Lower Fox Hills interval (Late Cretaceous: latest Campanian/Early Maastrichtian) in Badlands National Park area compared with the Type-areas of central South Dakota and eastern Wyoming [abs.]." The Geological Society of America, Boulder, Colorado, Rocky Mountain Section, Annual Meeting, Rapid City, South Dakota, *Program with Abstracts*, p. 39.

Stanton, T. W., 1888. "Paleontological notes [Fort Pierre fossils near Boulder, Colorado]." *Proceedings of the Colorado Scientific Society, Vol. 2*, pp. 184–187.

Stanton, T. W., 1925. "Well log in northern Ziebach County—the fossil content." *University of South Dakota Bulletin Series 25(14)*.

Stebinger, E., 1914. "The Montana Group of northwestern Montana." *U. S. Geological Survey Professional Paper 90-6*, pp. 61-68.

Steiner, M. B., and Shoemaker, E. M., 1996. "A hypothesized Manson Impact tsunami: Paleomagnetic and stratigraphic evidence in the Crow Creek Member, Pierre Shale." *The Geological Society of American Special Paper 302, Boulder, Colorado*, pp. 419–432.

Stephenson, L. W., 1941. "The larger invertebrate fossils of the Navarro group of Texas (exclusive of corals and crustaceans and exclusive of the fauna of the Escondido Formation)." *Texas University Publication 4101*, 641 p.

Stevenson, R. E., 1952. "Structures and stratigraphy of southwestern Butte County." *South Dakota State Geological Survey Report of Investigations 69*, 32 p.

Stoffer, P. W., and Chamberlain, J. A., Jr., 1996. "The Upper Pierre Shale/Lower Fox Hills Interval (Late Cretaceous; Latest Campanian/Early Maastrichtian) in Badlands National Park area compared with the Type-Areas of central South Dakota and eastern Wyoming [abs.]." The Geological Society of America, Boulder, Colorado, Rocky Mountain Section, Annual Meeting, Rapid City, South Dakota, *Program with Abstracts,* p. 14.

Tasch, P., 1973. *Paleobiology of the Invertebrates.* John Wiley & Sons, New York.

Teller-Marshall, S., and Bardack, D., 1978. "The morphology and relationships of the Cretaceous teleost *Apsopelix." Fieldiana Geology 41*, pp. 1–35.

Tourtelot, H. A., and Rye, R. O., 1969. "Distribution of oxygen and carbon isotopes in fossils of Late Cretaceous Age, Western Interior region of North America." *Geological Society of American Bulletin, Vol. 80*, pp. 1903–1922.

Tovell, W. M., 1951. "Geology of the Pembina Valley — Deadhorse Creek area." *Manitoba Department of Mines and Natural Resources Preliminary Report 47-7,* 7 p.

Tyrrell, J. B., 1890. "The Cretaceous of Manitoba." *American Journal of Science, 3rd Series, Vol. 40,* pp. 227–232.

Waage, K. M., 1965. "The Late Cretaceous coleoid cephalopod *Actinosepia canadensis* Whiteaves." *Peabody Museum of Natural History, Yale University Postilla 94,* 33 p.

Waage, K. M., 1968. "The type Fox Hills Formation Cretaceous (Maestrichtian), South Dakota, Part 1. Stratigraphy and Paleoenvironments." *Peabody Museum of Natural History, Yale University Bulletin 27.*

Ward, P. D., 1987. *The Natural History of Nautilus.* Allen and Unwin, Inc., Winchester, Massachusetts, 267 p.

Warren, P. S., 1930: "Three new ammonites from the Cretaceous of Alberta." *Royal Society of Canada, Transactions, Vol. 4(24),* pp. 21–26, pls. I–III.

Warren, P. S., 1931. "Invertebrate paleontology of the southern plains of Alberta." *American Association of Petroleum Geologists Bulletin, Vol. 15(10),* pp. 1283–1291, 3 pls.

Warren, P. S., 1934. "Paleontology of the Bearpaw Formation." *Royal Society of Canada, Transactions, Series 3, Section IV, No. 28,* pp. 81–97.

Weimer, R. J., 1960. "Upper Cretaceous stratigraphy, Rocky Mountain area." *American Association of Petroleum Geologists Bulletin, Vol. 44(1),* pp. 1–21.

Weimer, R. J., 1984. "Relation of unconformities, tectonics, and sea-level changes, Cretaceous of Western Interior, U.S.A.," *in* Schlee, J. S., ed., *Interregional Unconformities and Hydrocarbon Accumulation. American Association of Petroleum Geologists Memoir 36,* pp. 7–35.

Welles, S. P., 1943: "Elasmosaurid plesiosaurs with description of new material from California and Colorado." *Memoirs of the University of California, Vol. 13(3),* pp. 125–254.

Welles, S. P., 1952. "A review of the North American Cretaceous elasmosaurs." *University of California Publications in Geological Sciences 29,* pp. 47–144.

Welles, S. P., 1962. "A new species of elasmosaur from the Aptian of Colombia and a review of the Cretaceous plesiosaurs." *University of California Publications in Geological Sciences 44,* pp. 1–96.

Welles, S. P., and Bump, J. D., 1949. "*Alzadasaurus pembertoni,* a new elasmosaur from the Upper Cretaceous of South Dakota." *Journal of Paleontology 23,* pp. 521–535.

Westermann, G. E. G., 1971. "Form, structure and function of shell and siphuncle in coiled Mesozoic ammonoids." *Life Sciences Contributions, Royal Ontario Museum No. 78*, 39 p.

Wetzel, W., 1969. "Séltene wohnkammerinhalte von neoammoniten." *N. Jb. Geologica Paläontologie Mh.*, pp. 46–53.

Whiteaves, J. F., 1885. "Report on the invertebrata of the Laramie and Cretaceous rocks of the vicinity of the Bow and Belly Rivers and adjacent localities in the Northwest Territory." *Contributions to Canadian Palæontology, Vol. 1*, pp. 151–196.

Whiteaves, J. F., 1889. "On some Cretaceous fossils from British Columbia, the Northwest Territory and Manitoba." *Contributions to Canadian Palæontology, Vol. 1*, pp. 151–196.

Whiteaves, J. F., 1897. "On some remains of a *Sepia*-like cuttlefish from the Cretaceous rocks of the south Saskatchewan." *Canadian Rec. Science, Vol. 7*, pp. 459–460, pl. 2.

Whitfield, R. P., 1877. "Preliminary report on the paleontology of fossils from the Potsdam, Jurassic, and Cretaceous formations of the Black Hills of Dakota." *U.S. Geographical and Geological Survey Rocky Mountain Region Report (Powell)*, 49 p.

Whitfield, R. P., 1880. "Paleontology of the Black Hills of Dakota," *in* Newton, H., and Jenney, W. P., eds., *Report on the Geology and Resources of the Black Hills of Dakota.*" *U.S. Geographical and Geological Survey Rocky Mountain Region Report (Powell)*, pp. 325–468, 16 pls.

Whitfield, R. P., 1901. "Note on a very fine example of *Helicoceras stevensoni* preserving the outer chamber." *American Museum of Natural History Bulletin, Vol. XIV(XVI)*, pp. 219.

Whitfield, R. P., 1902. "Observations on and amended description of *Heteroceras simplicostatum* Whitfield." *American Museum of Natural History Bulletin, Vol 16(5)*, pp. 67–72, pls. 23–27.

Whitfield, R. P., 1907. "Notice of an American species of the genus *Hoploparia* McCoy from the Cretaceous of Montana." *Bulletin of the American Museum of Natural History 23*, pp. 459–462, pls 36.

Wiedmann, J., 1965. "Origin, limits, and systematic position of *Scaphites*." *Palæontology, Vol. 8(3)*, pp. 397–453.

Wieland, G. , R., 1896. "*Archelon ischyros*: a new gigantic cryptodire testudinate from the Fort Pierre Cretaceous of South Dakota." *American Journal of Science 2*, pp. 399–412.

Williams, G. D., and Bayliss, P., 1988. "Mineralogy of the Cretaceous shales in southeastern Saskatchewan." *Bulletin of Canadian Petroleum Geology, Vol. 36*, pp. 145–157.

Williams, G. D., and Burke, C. F. Jr., 1964. "Upper Cretaceous," *in* McCrossan, R. G., ed., *Geologic History of Western Canada: Calgary, Alberta.* Alberta Society of Petroleum Geologists, pp. 169–189.

Williams, G. D., and Moore, P. R., 1987. "The siliceous late Campanian Odanah Shale in southwestern Manitoba," *in* McNeil, D. H., ed., *Stratigraphy of Albian to Campanian Rocks in the Manitoba Escarpment, Pembina Mountain to Riding Mountain. Geological Association of Canada, field trip guidebook, trip 2: Saskatoon, Saskatchewan*, Geological Association of Canada, Waterloo, Ontario, pp. 37–57.

Williams, G. D., and Stelck, C. R., 1975. "Speculations on the Cretaceous palæogeography of North America." *The Geological Association of Canada Special Paper Number 13, Waterloo, Ontario*, pp. 6–17, fig. 8.

Williams, M. Y., 1930. "New species of marine invertebrate fossils from the Bearpaw Formation of southern Alberta." *Canadian National Museum Bulletin 63*, pp. 1–6.

Williston, S. W., 1903. "North American plesiosaurs, Part I." *Field Columbian Museum, Publication 73, Geo. Series, Vol. II(I).* 77 p.

Wing, M. E., 1940. "Bentonites of the Belle Fourche District." *South Dakota State Geological Survey Report of Investigations 35*, 29 p.

Witzke, B. J., Hammond, R. H., and Anderson, R. R., 1996. "Deposition of the Crow Creek Member, Campanian, South Dakota and Nebraska." *The Geological Society of America Special Paper 302, Boulder, Colorado*, pp. 433–456.

Witzke, B. J., Ludvigson, G. A., Poppe, J. R., and Raun, R. L., 1983. "Cretaceous paleogeography along the eastern margin of the Western Interior Seaway, Iowa, southern Minnesota, and eastern Nebraska and South Dakota," *in* Reynolds, M. W., and Dolly, E. D., eds., *Mesozoic Paleogeography of West-Central United States.* Rocky Mountain Section S.E.P.M., Boulder, Colorado, pp. 225–252.

Wright, C. W., 1957. "Superfamily Turrilitacea, p. L214-L228" *in* Arkell, W. J., Kummel, B., and Wright, C. W., "Mesozoic Ammonoidea, Pt. L, Mollusca 4," *in* Moore, R. C., ed., *Treatise on Invertebrate Paleontology.* University of Kansas Press, Lawrence, and The Geological Society of America, Boulder, Colorado, 490 p.

Wright, C. W., Calloman, J. H., and Howarth, M. K., 1996. *Treatise on Invertebrate Paleontology. Mollusca 4, Part L., revised, Vol 4: Cretaceous Ammonoidea.* The Geological Society of America, Boulder, Colorado, and The University of Kansas Press, Lawrence, 511 p.

Wright, C. W., and Wright, E. V., 1951. "A survey of the fossil cephalopoda of the Chalk of Great Britain." *Palaeontographical Society (London) Monograph*, 40 p.

Wright, E. K., 1987. "Stratification and paleocirculation of the Late Cretaceous Western Interior Seaway of North America." *The Geological Society of America Bulletin, Boulder, Colorado, Vol. 99*, pp. 480–490.

Young, K., 1963. "Upper Cretaceous ammonites from the Gulf Coast of the United States." *Texas University Publication 6304*, 373 p., 82 pls.

Young, H. R., and Bristol, C. C., 1982. "Geochemistry and mineralogy of the siliceous Odanah Shale (Cretaceous) of southwestern Manitoba [abs.]." Geological Association of Canada Annual Meeting, Winnipeg, Manitoba, *Program with Abstracts*, p. 88.

Young, H. R., and Moore, P. R., 1987. "The siliceous late Campanian Odanah Shale in southwestern Manitoba," *in* McNeil, D. H., ed., *Stratigraphy of Albian to Campanian Rocks in the Manitoba Escarpment, Pembina Mountain to Riding Mountain. Geological Association of Canada, Field Trip Guidebook, Trip 2, Saskatoon, Saskatchewan*, pp. 37–57.

Young, H. R., and Moore, P. R., 1994. "Composition and depositional environment of the siliceous Odanah Member (Campanian) of the Pierre Shale in Manitoba," *in* Shurr, G. W., Ludvigson, G. A., and Hammond, R. H., eds., *Perspectives on the Eastern Margin of the Cretaceous Western Interior Basin. The Geological Society of America Special Paper 287, Boulder, Colorado*, pp. 175–196.

Zangerl, R., 1953a. "The vertebrate fauna of the Selma Formation of Alabama, part III—The turtles of the Family Protostegidae." *Fieldiana Geology Memoirs 3*, pp. 57–134.

Zangerl, R., 1953b. "The vertebrate fauna of the Selma Formation of Alabama, part IV—The turtles of the Family Toxochelyidae." *Fieldiana Geology Memoirs 3*, pp. 135–278.

Zapp, A. D., and Cobban, W. A., 1960. "Some Late Cretaceous strand lines in northwestern Colorado and northeastern Utah," *in Short Papers in the Geological Sciences. U.S. Geological Survey Professional Paper 400-B*, pp. B246–249.

Zapp, A. D., and Cobban, W. A., 1962. "Some Late Cretaceous strand lines in southern Wyoming," *in Short Papers in Geology, Hydrology, and Topography. U.S. Geological Survey Professional Paper 450-D*, pp. 52D–D55.

Ziegler, A. M., and Hulver, M. L., 1989. "Marine connections between the Western Interior Seaway and the Labrador Sea." The Geological Society of America, Boulder, Colorado, *Abstracts with Programs, Vol. 21*, p. A168.

Zittel, K. A., 1884. "Handbuch der palæontologie." *1, Abt 2, Lief, 3, Cephalopoda, R. Oldenbourg*, pp. 329–522.

Zittel, K. A., 1895. "Grundzüge der paläeontologie (palaeozoologie)." *R. Oldenbourg*, 971 p.

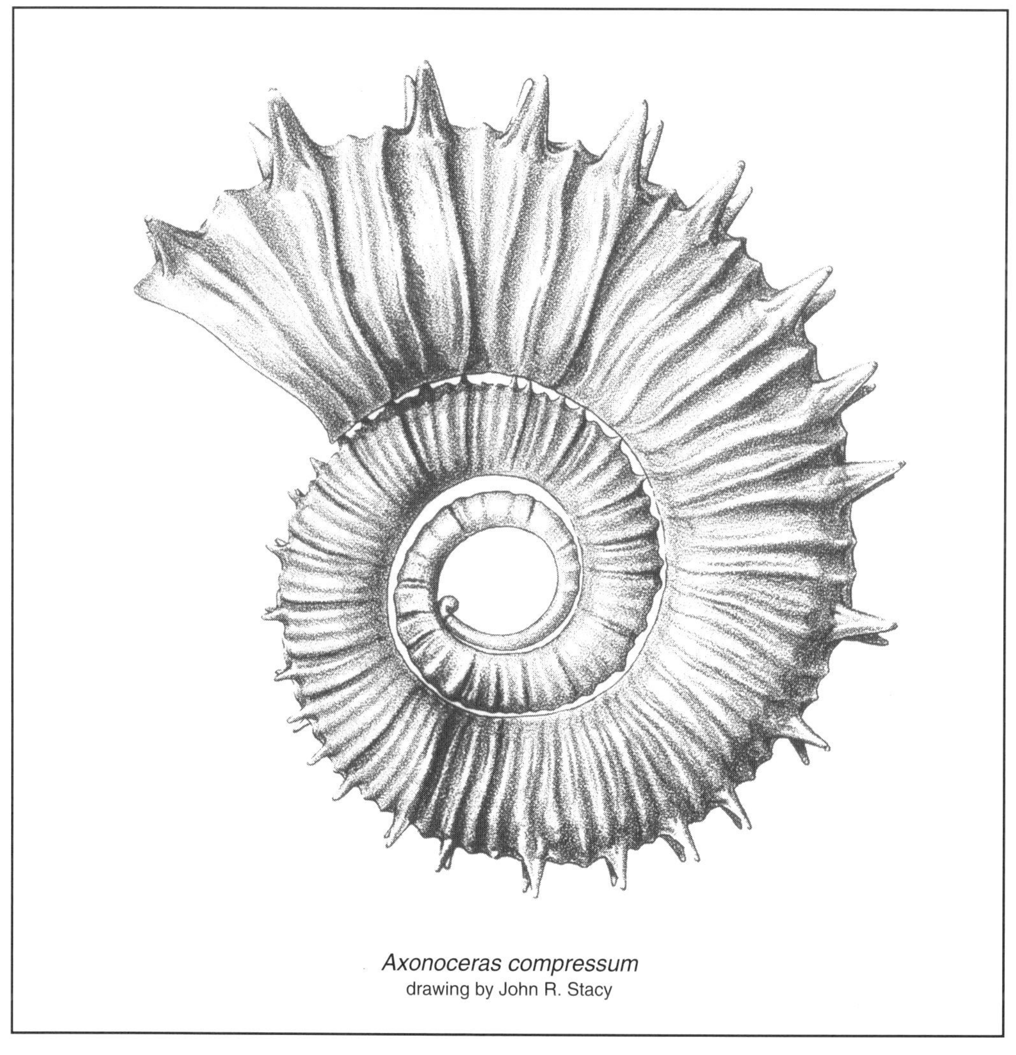

Axonoceras compressum
drawing by John R. Stacy

* Parentheses enclosing the author's name indicate that his original description placed the species in a genus not currently accepted for that species.

The Known Range of the Ammonite Species

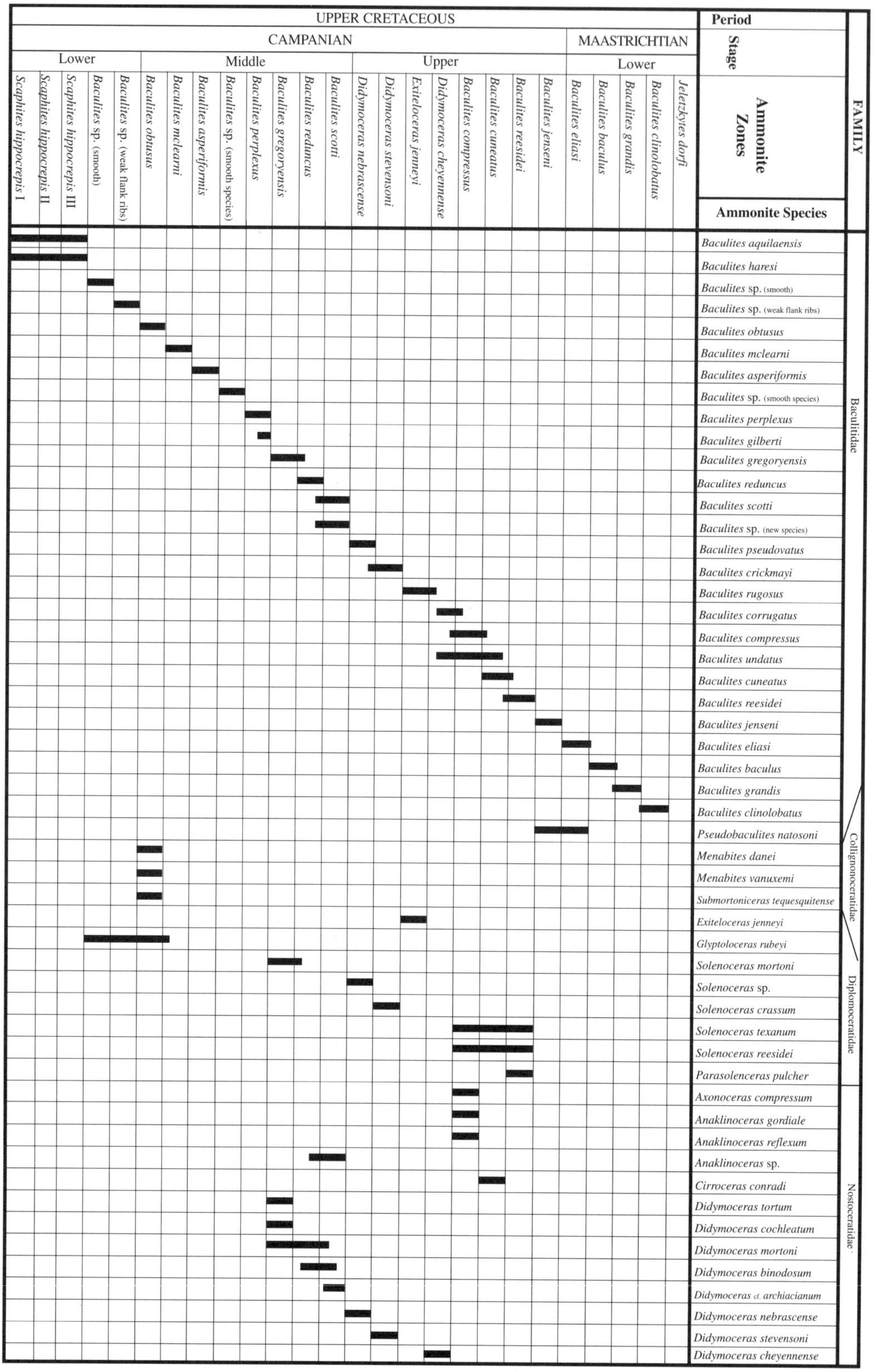

Period: Upper Cretaceous

Stage: Campanian (Lower, Middle, Upper) — Maastrichtian (Lower)

Ammonite Zones (Z1–Z26):

Zone	Name	Stage
Z1	Scaphites hippocrepis I	Campanian – Lower
Z2	Scaphites hippocrepis II	Campanian – Lower
Z3	Scaphites hippocrepis III	Campanian – Lower
Z4	Baculites sp. (smooth)	Campanian – Lower
Z5	Baculites sp. (weak flank ribs)	Campanian – Lower
Z6	Baculites obtusus	Campanian – Middle
Z7	Baculites mclearni	Campanian – Middle
Z8	Baculites asperiformis	Campanian – Middle
Z9	Baculites sp. (smooth species)	Campanian – Middle
Z10	Baculites perplexus	Campanian – Middle
Z11	Baculites gregoryensis	Campanian – Middle
Z12	Baculites reduncus	Campanian – Middle
Z13	Baculites scotti	Campanian – Middle
Z14	Didymoceras nebrascense	Campanian – Upper
Z15	Didymoceras stevensoni	Campanian – Upper
Z16	Exiteloceras jenneyi	Campanian – Upper
Z17	Didymoceras cheyennense	Campanian – Upper
Z18	Baculites compressus	Campanian – Upper
Z19	Baculites cuneatus	Campanian – Upper
Z20	Baculites reesidei	Campanian – Upper
Z21	Baculites jenseni	Campanian – Upper
Z22	Baculites eliasi	Maastrichtian – Lower
Z23	Baculites baculus	Maastrichtian – Lower
Z24	Baculites grandis	Maastrichtian – Lower
Z25	Baculites clinolobatus	Maastrichtian – Lower
Z26	Jeletzkytes dorfi	Maastrichtian – Lower

Ammonite Species — range across zones (█ = known range):

Ammonite Species	Family	Z1	Z2	Z3	Z4	Z5	Z6	Z7	Z8	Z9	Z10	Z11	Z12	Z13	Z14	Z15	Z16	Z17	Z18	Z19	Z20	Z21	Z22	Z23	Z24	Z25	Z26
Baculites aquilaensis	Baculitidae	█	█	█																							
Baculites haresi	Baculitidae	█	█	█																							
Baculites sp. (smooth)	Baculitidae				█																						
Baculites sp. (weak flank ribs)	Baculitidae					█																					
Baculites obtusus	Baculitidae						█																				
Baculites mclearni	Baculitidae							█																			
Baculites asperiformis	Baculitidae								█																		
Baculites sp. (smooth species)	Baculitidae									█																	
Baculites perplexus	Baculitidae										█																
Baculites gilberti	Baculitidae										█																
Baculites gregoryensis	Baculitidae											█															
Baculites reduncus	Baculitidae												█														
Baculites scotti	Baculitidae													█													
Baculites sp. (new species)	Baculitidae														█												
Baculites pseudovatus	Baculitidae															█											
Baculites crickmayi	Baculitidae																█										
Baculites rugosus	Baculitidae																	█									
Baculites corrugatus	Baculitidae																	█									
Baculites compressus	Baculitidae																	█	█								
Baculites undatus	Baculitidae																		█	█							
Baculites cuneatus	Baculitidae																			█							
Baculites reesidei	Baculitidae																				█						
Baculites jenseni	Baculitidae																					█					
Baculites eliasi	Baculitidae																						█				
Baculites baculus	Baculitidae																							█			
Baculites grandis	Baculitidae																								█	█	
Baculites clinolobatus	Baculitidae																									█	█
Pseudobaculites natosoni	Collignoniceratidae																				█	█					
Menabites danei	Collignoniceratidae						█																				
Menabites vanuxemi	Collignoniceratidae						█																				
Submortoniceras tequesquitense	Collignoniceratidae						█																				
Exiteloceras jenneyi	Diplomoceratidae																█										
Glyptoloceras rubeyi	Diplomoceratidae				█	█	█	█																			
Solenoceras mortoni	Diplomoceratidae											█															
Solenoceras sp.	Diplomoceratidae													█													
Solenoceras crassum	Diplomoceratidae															█											
Solenoceras texanum	Diplomoceratidae																	█	█	█	█						
Solenoceras reesidei	Diplomoceratidae																	█	█	█	█						
Parasolenceras pulcher	Diplomoceratidae																			█	█						
Axonoceras compressum	Diplomoceratidae																	█	█								
Anaklinoceras gordiale	Diplomoceratidae																	█	█								
Anaklinoceras reflexum	Diplomoceratidae																	█	█								
Anaklinoceras sp.	Diplomoceratidae																		█	█							
Cirroceras conradi	Nostoceratidae												█	█													
Didymoceras tortum	Nostoceratidae											█															
Didymoceras cochleatum	Nostoceratidae											█															
Didymoceras mortoni	Nostoceratidae											█	█	█													
Didymoceras binodosum	Nostoceratidae												█	█													
Didymoceras cf. archiacianum	Nostoceratidae													█													
Didymoceras nebrascense	Nostoceratidae														█												
Didymoceras stevensoni	Nostoceratidae															█											
Didymoceras cheyennense	Nostoceratidae																█	█									

     Ammonites and the other Cephalopods of the Pierre Seaway   Illustration by N. Larson, 1996. Computer Graphics by M. Zenker.

The Known Range of the Ammonite Species

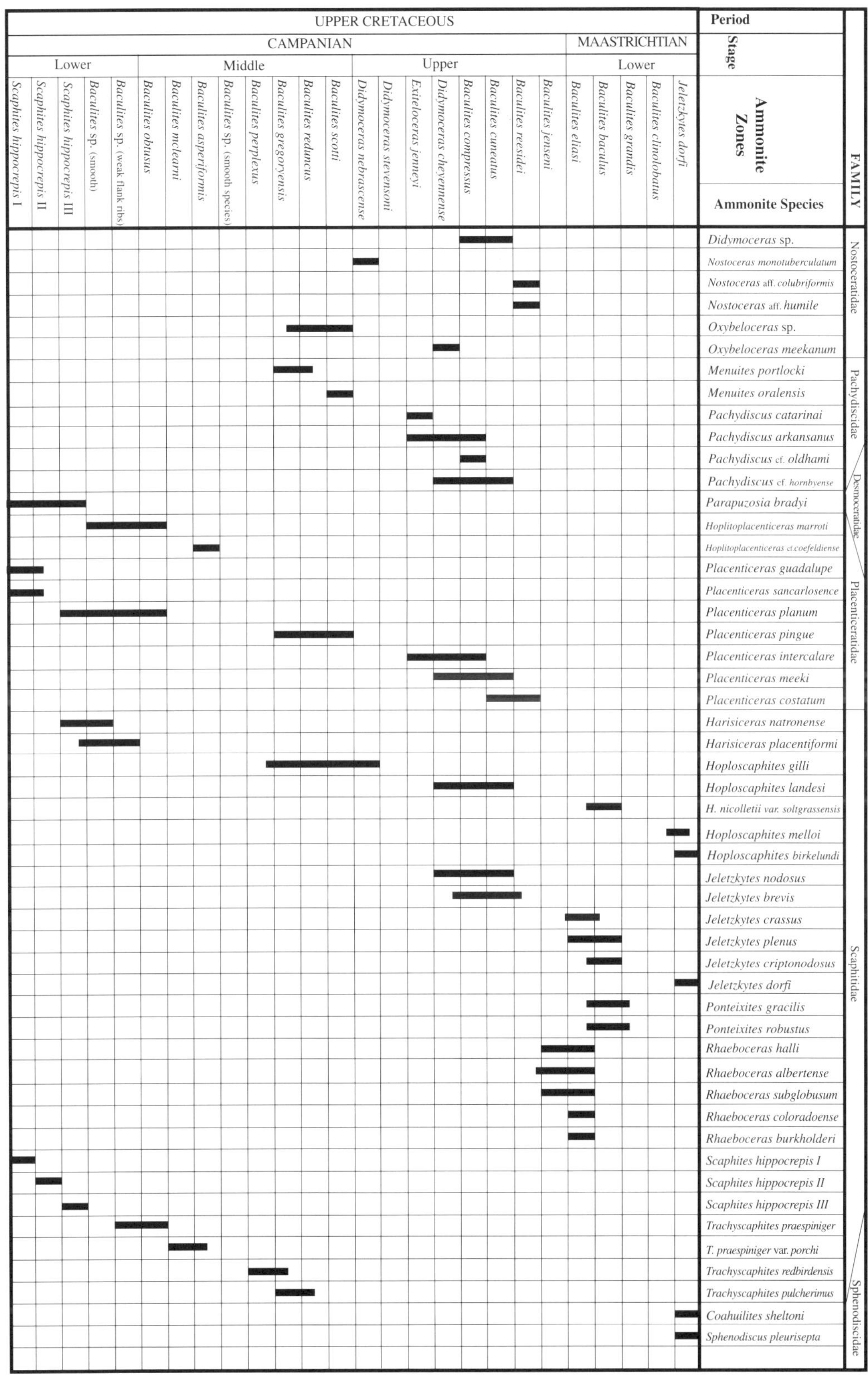

Illustration by N. Larson, 1996. Computer Graphics by M. Zenker.

145

More than three years of research and effort have gone into this book in our attempt to ensure that it is a complete, accurate, informative, and readable guide. The last 150 years of collecting and research has taught us a great deal about the seaway that covered a large portion of the Western Interior of North America during most of the Cretaceous period and the life that inhabited it. We anticipate that future collecting and research will reveal even more about that past.

As we finally go to press, there are numerous new ammonite descriptions in process and many more ammonites that will take years of compiling and examination before their descriptions are written and published. Some works to watch for are those dealing with the genera within the family, Scaphitidae, Baculitidae, and Nostoceratidae, by W. J. Kennedy and W. A. Cobban. These two scientists in collaboration with N. H. Landman will soon be publishing papers on the *Sphenodiscus* and *Coahuilites* of the Western Interior. Landman, Kennedy, Cobban, and I are attempting to clarify and name the numerous undescribed species of scaphites in the genera of *Hoploscaphites* and *Jeletzkytes* from the Pierre Shale. Jim and Joyce Grier are studying and writing on two other genera of scaphites, the *Rhaeboceras* and *Ponteixites*.

This printing is by no means a complete and final version of this book. There are many other cephalopods that need descriptive work and names. You need only look at the Range Zone fossils from the Western Interior to see that there is much work yet to do. From *Baculites* sp. (smooth), *Baculites* sp. (weak flank ribs), and *Baculites* sp. (smooth species) to the *Tusoteuthis* and the *Eutrephoceras*, you can comprehend some of the problems that must be resolved before a nearly final version of this book can be published. That will involve many more years of research.

The future holds great promise for new discoveries and research in ammonite biology. Currently there are fewer paleontologists studying fossil cephalopods than probably at any other time during the twentieth century. We are fortunate to have so much data available from past researchers to assist us in understanding these remarkable animals. These early paleontologists began an important study that we must continue. Our hope is that this book will help generate a widespread fascination in these animals. As more people become interested in collecting and studying cephalopods, the frontiers of paleontology will expand. With new and better preparation techniques, our understanding of just how unique these marine animals were will unfold.

If this book helps you with identification of your cephalopods or provides any references that were previously unknown to you, then our reasons for compiling this body of information and writing

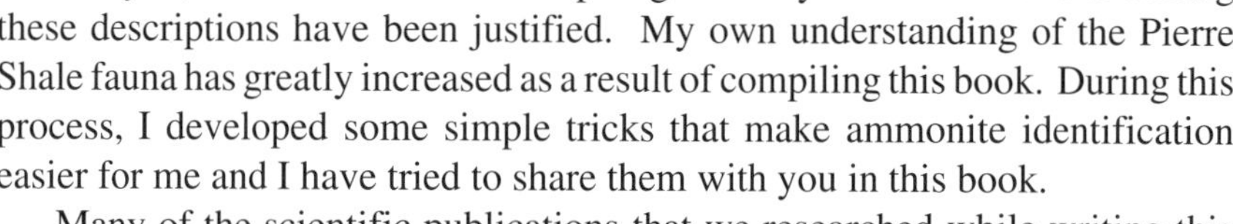

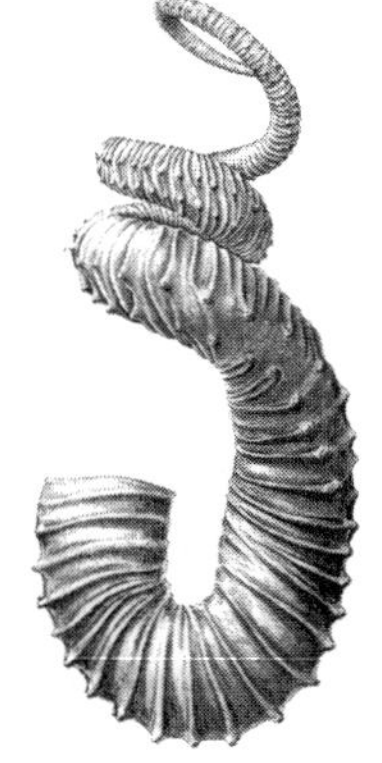

these descriptions have been justified. My own understanding of the Pierre Shale fauna has greatly increased as a result of compiling this book. During this process, I developed some simple tricks that make ammonite identification easier for me and I have tried to share them with you in this book.

Many of the scientific publications that we researched while writing this manuscript contained errors. We have hopefully corrected those errors here, and have not added too many of our own. It is with great pleasure that we present you with this work. Our hope is that it will become an invaluable reference guide as you collect in the field or attempt to identify the specimens already in your collections.

Neal L. Larson

ABOUT BLACK HILLS INSTITUTE OF GEOLOGICAL RESEARCH, INC.

The Black Hills Institute of Geological Research, Inc., is an earth science supply house and fossil preparatory that supplies mineral and fossil specimens to museums, schools, scientists, businesses, and individuals. The Black Hills Institute's business offices, preparatory and exhibit facilities are located on Main Street in the historic mining town of Hill City in the Black Hills of South Dakota. In 1973, the company was organized as Black Hills Minerals, and was incorporated as Black Hills Institute of Geological Research, Inc., in 1978. Today, Black Hills Institute is the largest business of its type in the world.

Ammonites were some of the first fossils collected and prepared by institute staff. As the years passed, the institute began supplying a variety of fossils, including gastropods, cephalopods, trilobites, echinoderms, crustaceans, and other invertebrate fossils to clients. Vertebrate fossils also found their way into the expanding inventory, eventually resulting in our ability to offer mounted, original, and cast skeletons of reptiles, mammals, birds, and even dinosaurs for exhibition and research. Today, fossils supplied by the institute are displayed in major museums in the United States, Canada, Peru, Japan, China, Wales, Ireland, the Netherlands, Germany, Austria, France, Switzerland, Italy, Australia, and Mexico.

Despite this growth, the institute's owners have maintained a romance with ammonites. Since our beginnings, we have accummulated one of the largest and most comprehensive collections of Late Cretaceous ammonites anywhere in the world. It is part of this collection that provided the basis for this book, *Ammonites and the Other Cephalopods of the Pierre Seaway*. Specimens from this collection are now on display at the Black Hills Museum's temporary site.

Peter L. Larson, President and Founder

ABOUT THE BLACK HILLS MUSEUM OF NATURAL HISTORY

The birth of the Black Hills Museum of Natural History was first publicly announced in September of 1990, but its conception occurred in the late 1950s. It was during our childhood that Neal and I began dreaming of creating a museum to house and display our mineral and fossil discoveries. It was that dream, with the support and help of family, friends, and other dedicated collectors, that eventually led to the opening of the museum's first exhibits in the main building of the Black Hills Institute of Geological Research in Hill City, South Dakota, in the summer of 1992. Since 1992, the exhibits have multiplied to fill the display space beyond capacity, and traveling exhibits visit local schools. In addition, the museum sponsors an annual Natural History Days Festival every spring.

The Black Hills Museum of Natural History is purchasing 18 acres of forest and meadow on a hill overlooking the late 1800s mining town of Hill City, South Dakota, in the heart of the Black Hills. It is here that a large, permanent facility will be constructed. The museum, once built, will offer exhibits on history, geology, mineralogy, and paleontology. There will also be lecture halls, seasonal and traveling exhibits, and even a theater. In addition, the large and comprehensive ammonite collection assembled by the Black Hills Institute will be permanently housed in the Black Hills Museum of Natural History. An excellent reference collection of ammonites at Black Hills Institute is even now available to those engaged in research projects.

Peter L. Larson, Vice President and Founder

Neal L. Larson is a South Dakota native and has been interested in ammonites since childhood. He loves all ammonites. His passion has been collecting, studying, and identifying Campanian and Maastrichtian Age ammonites from the Western Interior of North America. An increasing collection of ammonites from other sites has led to a growing interest in ammonites from around the world. Neal served as a content specialist for Glenco Publishing in preparation of a ninth- and tenth-grade level text titled, *Biology, The Dynamics of Life*. Presently, he is coauthoring a publication for the American Museum of Natural History on *Scaphites of the Pierre Shale* with Dr. N. H. Landman, Dr. W. A. Cobban, and Dr. W. J. Kennedy. For twenty-plus years, Neal has been vice president of the Black Hills Institute of Geological Research, Inc., as well as a fossil collector and preparator with the institute. As president of the Black Hills Museum of Natural History, he has spearheaded the efforts to purchase land for the museum site and establish an annual Natural History Days Festival in Hill City, South Dakota.

Steven D. Jorgensen grew up near Yankton, and now resides in Sioux Falls, South Dakota. He has collected and studied ammonites for more than 25 years. Steve's primary interest is Upper Cretaceous ammonites, from the Turonian and Campanian Ages of the Western Interior of North America. He is also an Engineer-in-Training (EIT), a Professional Geologist (#130) in the State of Wyoming, and a Certified Professional Geologist (#9596) by the American Institute of Professional Geologists. His current research includes a nearly complete publication comparing the Carlile Shale of the Black Hills to the Carlile Shale in eastern South Dakota, with particular attention to the shark and ammonite taxa similarities and differences.

Robert A. Farrar, a native of the Cleveland, Ohio, area, developed his interest in fossils in junior high. Robert moved to South Dakota in 1973 to attend school, meeting Neal and Peter Larson while all were college students. He has served as secretary/treasurer for the Black Hills Institute of Geological Research, Inc., since its incorporation in 1978, and is presently the institute's CEO as well. His previous publication credits include coauthorship with Peter L. Larson of *What Is a Fossil?* and *Fossil Vertebrates of the White River Badlands*. Robert's current research involves the study of Upper Cretaceous, Cenomanian Age, marine fishes of Lebanon and Pegmatite phosphate mineralogy. A major portion of the bibliography in this publication is the result of his exhaustive search of the literature.

Peter L. Larson has shared with his brother, Neal, a lifelong love for, and fascination with, ammonites. Together, they have amassed the magnificent collection of prepared and unprepared ammonite specimens presently owned by the Black Hills Institute of Geological Research, Inc. Peter is the chief author of the institute's *What Is It?* series of educational posters and pamphlets including *What Is an Ammonite?* He is also the primary author of the institute's publications titled *Fossil Vertebrates of the White River Badlands* and *The Pierre Shale and Its Macrofauna*. *The Pierre Shale and Its Macrofauna,* no longer in print, is the precursor to the present publication, *Ammonites and the Other Cephalopods of the Pierre Seaway*. Peter is well known for his research papers and popular articles on *Tyrannosaurus rex* and his current research projects include a book-in-progress on that tyrant king of the Cretaceous. President and founder of Black Hills Institute of Geological Research, Inc., he is also vice president of the Black Hills Museum of Natural History in Hill City, South Dakota.